T0093450

Recursive Filtering for 2-D Shift-Varying Systems with Communication Constraints

Recursive Filtering for 2-D Shift-Varying Systems with Communication Constraints

Jinling Liang
Zidong Wang
Fan Wang

CRC Press
Taylor & Francis Group
Boca Raton London New York

CRC Press is an imprint of the
Taylor & Francis Group, an **informa** business
A CHAPMAN & HALL BOOK

First edition published 2022
by CRC Press
6000 Broken Sound Parkway NW, Suite 300, Boca Raton, FL 33487-2742

and by CRC Press
2 Park Square, Milton Park, Abingdon, Oxon, OX14 4RN

© 2022 Jinling Liang, Zidong Wang and Fan Wang

CRC Press is an imprint of Taylor & Francis Group, LLC

ISBN: 978-1-032-03817-9 (hbk)
ISBN: 978-1-032-03822-3 (pbk)
ISBN: 978-1-003-18921-3 (ebk)

Typeset in Latin Modern font
by KnowledgeWorks Global Ltd.

To our families.

Contents

Preface

Two-dimensional (2-D) systems have been receiving a steadily growing research interest for their promising application insights in various engineering fields including image processing, electricity transmission, chemical processes, and multi-variable networks. In contrast with the traditional 1-D systems whose state evolves along a single direction, the information in 2-D systems propagates along two independent directions, thereby capable of modeling many real-world systems. It is crucial to reconstruct/estimate the system states of interest. Although considerable attention has been paid to 2-D filtering issues, the counterparts for 2-D shift-varying systems have been greatly neglected, where the conventional filtering techniques for shift-invariant systems are inapplicable anymore. On the other hand, communication constraints are inevitable for systems over communication networks, which may result in a variety of undesirable phenomena including measurement degradations, sensor delays, signal quantization, and so forth. All these phenomena play an important role in estimating the true states and have a great influence on the filtering performance. The traditional filtering methods have limitations in handling the filtering problems for 2-D shift-varying systems with communication constraints, and new recursive filtering strategies are of urgency to be developed, which motivates the current research.

The primary objective of this book is to present the up-to-date research developments and novel methodologies regarding recursive filtering for 2-D shift-varying systems with varieties of communication constraints. A systematic investigation on recursive filter/estimator design and performance analysis has been developed in the 2-D framework. A combination of the intensive stochastic analysis, recursive Riccati-like equations, variance-constrained approach, and matrix decomposition technique is utilized to subtly design the filter gains and expound effects of communication constraints on the filtering performance. Moreover, this book provides valuable reference materials for researchers who wish to explore the area of 2-D filtering issues.

The compendious frame and description of the book are given as follows. Chapter 1 introduces the recent progress on the filtering problems for 2-D systems and the outline of the book. Chapter 2 deals with the recursive minimum-variance filtering problem for a class of 2-D shift-varying systems with stochastic nonlinearity and degraded measurements. The robust Kalman filtering problem is investigated in Chapter 3 for a class of 2-D uncertain systems with both additive and multiplicative noises, where the norm-bounded parameter uncertainties enter into both the state and the output matrices.

In Chapter 4, the robust finite-horizon filter design problem is discussed for 2-D shift-varying uncertain systems with incomplete measurements that cover randomly occurring sensor delays and missing measurements in a unified form. Chapter 5 addresses the recursive filtering problem for a class of 2-D systems subjected to missing measurements, whose occurrences are depicted by a series of uncorrelated stochastic variables obeying individual Bernoulli distributions with uncertain probabilities. Chapter 6 copes with the resilient filtering problem for 2-D shift-varying systems with redundant channels and gain perturbations. In Chapter 7, the distributed filtering problem is considered for 2-D shift-varying systems under random access protocol. For a class of linear shift-varying repetitive processes with communication constraints, Chapter 8 is concerned with the recursive filtering problem with uniform quantization and Round-Robin protocol. Chapter 9 discusses the recursive filtering problem for linear shift-varying repetitive processes with an event-triggering mechanism. Chapter 10 gives the conclusion and some possible future research topics.

This book is a research monograph whose intended audiences are graduate and postgraduate students as well as researchers.

Acknowledgments

We would like to acknowledge the help of many people who have been directly involved in various aspects of the research leading to this book. Special thanks go to Prof. Xiaohui Liu from Brunel University London, Uxbridge, United Kingdom. Finally, we would like to thank the editors at CRC Press for their professional and efficient handling of this monograph.

The writing of this book was supported in part by the National Key Research and Development Program of China under Grant 2018AAA0100202, and in part by the National Natural Science Foundation of China under Grants 61673110, 61873148, 61933007, and 61903082.

Author Biographies

Jinling Liang received the B.Sc. and M.Sc. degrees in mathematics from Northwest University, Xi'an, China, in 1997 and 1999, respectively, and the Ph.D. degree in applied mathematics from Southeast University, Nanjing, China, in 2006. She was a Post-Doctoral Research Fellow from April 2007 to March 2008 and a Visiting Research Fellow from January to March 2010, with the Department of Information Systems and Computing, Brunel University, London, UK, sponsored by the Royal Society, UK. From March to August 2009, she was a Research Associate with The University of Hong Kong, Hong Kong. From January to March 2017, she was a Temporary Associate Research Scientist with the Texas A&M University at Qatar, Qatar. From January to April 2018, she was a Senior Research Associate with the City University of Hong Kong, Kowloon, Hong Kong. She is currently a Professor in the School of Mathematics, Southeast University.

Professor Liang's research interests include 2-D systems, stochastic systems, complex networks, robust filtering, and bioinformatics. She has published around 90 papers in refereed international journals. According to the Web of Science, her publications have received more than 3,000 citations with h-index 40. She is currently serving or has served as an Associate Editor for *IEEE Transactions on Neural Networks and Learning Systems, International Journal of Computer Mathematics, IET Control Theory & Applications, International Journal of Systems Science,* and *Neurocomputing.* She is also a member of the program committees of more than 20 international conferences, and serves as a very active reviewer for many international journals.

Zidong Wang is currently a Professor of Department of Computer Science at Brunel University London in the United Kingdom. From January 1997 to December 1998, he was an Alexander von Humboldt Research Fellow with the Control Engineering Laboratory, Ruhr-University Bochum, Germany. From January 1999 to February 2001, he was a Lecturer with the Department of Mathematics, University of Kaiserslautern, Germany. From March 2001 to July 2002, he was a University Senior Research Fellow with the School of Mathematical and Information Sciences, Coventry University, UK. In August 2002, he joined the Department of Information Systems and Computing, Brunel University, UK, as a Lecturer, and was then promoted to a Reader in September 2003 and to a Chair Professor in July 2007.

Professor Wang's research interests include dynamical systems, signal processing, bioinformatics, control theory, and applications. He has published more than 200 papers in refereed international journals. According to the Web of Science, his publications have received more than 31,366 citations (excluding self-citations) with h-index 106. He was awarded the Humboldt research fellowship in 1996 from Alexander von Humboldt Foundation, the JSPS Research Fellowship in 1998 from Japan Society for the Promotion of Science, and the William Mong Visiting Research Fellowship in 2002 from The University of Hong Kong.

Professor Wang is currently serving or has served as the Editor-in-Chief for *Neurocomputing*, the Editor-in-Chief for *International Journal of Systems Science,* an Action Editor for *Neural Networks,* an Associate Editor for 12 international journals including *IEEE Transactions on Automatic Control, IEEE Transactions on Neural Networks, IEEE Transactions on Signal Processing, IEEE Transactions on Systems, Man, and Cybernetics-Part C, IEEE Transactions on Control Systems Technology, Circuits, Systems & Signal Processing, Asian Journal of Control,* an Editorial Board Member for *Information Fusion, IET Control Theory & Applications, Complexity, International Journal of Systems Science, Neurocomputing, International Journal of General Systems, Studies in Autonomic, Data-driven and Industrial Computing,* and an Associate Editor on the Conference Editorial Board for the IEEE Control Systems Society.

Professor Wang is a Member of the Academia Europaea (section of Physics and Engineering Sciences), a Fellow of the IEEE (for contributions to networked control and complex networks), a Fellow of the Chinese Association of Automation, a Member of the IEEE Press Editorial Board, a Member of the EPSRC Peer Review College of the UK, a Fellow of the Royal Statistical Society, a member of program committee for many international conferences, and a very active reviewer for many international journals. He was nominated an appreciated reviewer for *IEEE Transactions on Signal Processing* in 2006–2008 and 2011, an appreciated reviewer for *IEEE Transactions on Intelligent Transportation Systems* in 2008; an outstanding reviewer for *IEEE Transactions on Automatic Control* in 2004 and for the journal *Automatica* in 2000.

Fan Wang received the B.Sc. degree in mathematics from Hefei Normal University in 2012, and the Ph.D. degree in applied mathematics from Southeast University, Nanjing, China, in 2018. From 2016 to 2018, she was a visiting Ph.D. student with the Department of Information Systems and Computing, Brunel University London, Uxbridge, UK. She was a Research Associate with the Department of Mechanical Engineering, The University of Hong Kong, Hong Kong, in 2019, for two months. She is currently a Postdoctoral Research Fellow with Southeast University, Nanjing, China.

Dr. Wang has published over 20 papers in refereed international journals. Her current research interests include stochastic systems, 2-D systems, time-varying systems, optimal control, and robust filtering. She is a very active reviewer for several international journals.

List of Figures

List of Tables

Symbols

$\mathbb{R}^n$ The n-dimensional Euclidean space.

$\mathbb{R}^{n \times m}$ The set of all $n \times m$ real matrices.

I The identity matrix with compatible dimensions.

0 The zero matrix with compatible dimensions.

$\mathbf{1_n}$ The n-dimensional vector with elements all being 1.

$\|\cdot\|$ The Euclidean norm of real vectors or the spectral norm of real matrices.

A^T The transpose matrix of A.

$\text{tr}\{A\}$ The trace of matrix A.

$A^{(kl)}$ The (k, l)-th entry of matrix A.

$\text{diag}_{1 \le i \le n}\{A_i\}$ The block diagonal matrix $\text{diag}\{A_1, A_2, \ldots, A_n\}$ with diagonal blocks being matrices $A_1, A_2, \ldots, A_n$.

$\text{col}_m\{v_l\}$ The vector $[v_1^T\ v_2^T \cdots v_m^T\ ]^T$.

M^{-1} The inverse matrix of a nonsingular matrix M.

$\lambda_{\max}(M)$ The maximum eigenvalue of a symmetric matrix M.

$X > Y$ The matrix $X - Y$ is positive definite, where X and Y are symmetric matrices.

$X \ge Y$ The matrix $X - Y$ is positive semi-definite, where X and Y are symmetric matrices.

$\otimes$ The Kronecker product of matrices.

$\circ$ The Hadamard product of matrices.

$\cup$ The union of certain sets.

$\text{Prob}\{\cdot\}$ The occurrence probability of event "$\cdot$".

$\mathbb{E}\{x\}$ The expectation of stochastic variable x.

$\mathbb{E}\{x|y\}$ The expectation of x conditional on y, where x and y are both stochastic variables.

$\text{Var}\{x\}$ The variance of stochastic variable x.

$\text{Cov}\{x, y\}$ The covariance of stochastic variables x and y.

$\delta(i, k)$ The Kronecker delta function with $\delta(i, k)$ being 1 for $i = k$ and zero otherwise.

$\{a_i\}_{i=1}^{m}$ The set $\{a_1, a_2, \ldots, a_m\}$.

$[a \ b]$ The set $\{a, a + 1, \ldots, b\}$, where both a and b are integers.

$[a \ \infty)$ The set $\{a, a + 1, \ldots\}$, where a is an integer.

$\mathscr{S}[a \ b]$ The set $\{(i, j) | i, j \in [a \ b]\}$.

$\text{mod}\,(a, b)$ The remainder on division of the integer a by the positive integer b.

1

Introduction

1.1 2-D Systems

Two-dimensional (2-D) systems arise primarily from the practical requirements for depicting the information broadcast of the target plants over two
directions. In contrast with the traditional one-dimensional (1-D) systems
whose states evolve along a single direction, the states of 2-D systems propagate along two independent directions leading to more complicated dynamics [58]. Thanks to the inherent feature of two-directional propagations, 2-D
systems have shown their promising applications in many engineering fields,
such as manufacturing, industrial automation, grid sensor networks, and environment monitoring.

It is worth noting that 2-D systems provide a powerful tool to describe
the system dynamics with multiple independent variables. Typical examples
of practical 2-D systems include, but are not limited to, iterative circuits,
batch processes, thermal processes, digital image processing, seismographic
data analysis, and multi-variable network visualization [14, 51, 62, 83, 123].
Particularly, unlike a unilateral or sequential circuit with one-directional evolution, signals of the bilateral circuit flow in two different directions. States
of the batch process transmit along both the time and batch directions in
industry. The reactor temperature of the heating process varies with different spatial and temporal positions. A general 2-D system, from a mathematical point of view, has a formulation of partial differential/difference
equations with evolutions of two-variable functions and diffusions, which
can represent the dynamical evolution with respect to the fluid, heat, and
electrodynamics. To date, both theoretical developments and practical applications of the 2-D systems have received considerable research interest.
Moreover, a number of basic concepts and theories have been developed for
various 2-D systems, which mainly embrace the controllability and observability issues [84, 91, 132, 181, 234], the model reduction or approximation
[35, 63, 90, 231, 236], the stability analysis   [48, 100, 133, 148, 152, 171], and
the control and filtering issues [47, 103, 108, 168, 176, 177, 229].

1.1.1 Some Classical 2-D Models

Introduction of the 2-D state-space models dates back to the 1970s. One of the earliest 2-D state-space models has been developed in [68] for a multi-dimensional linear iterative circuit, where the general response formula has been obtained on the strength of two-tuple powers of certain matrix. Based on a similar approach, a linear discrete state-space model has been generalized in [162] from the model defined in a single-dimensional time to that in a 2-D space, thereby being able to describe the linear image processing. The corresponding generalization contains the novel state transition matrix, observability, and controllability in the 2-D framework. Ever since then, some classical 2-D models have been proposed in the existing literature, and the relationships between them can be clearly stated [10, 57, 58, 83, 162].

Consider the Roesser model described as follows [162]:

$$
\begin{bmatrix} x^h(i+1,j) \\ x^v(i,j+1) \end{bmatrix} = \begin{bmatrix} A_{11} & A_{12} \\ A_{21} & A_{22} \end{bmatrix} \begin{bmatrix} x^h(i,j) \\ x^v(i,j) \end{bmatrix} + \begin{bmatrix} B_1 \\ B_2 \end{bmatrix} u(i,j), \qquad (1.1)
$$

where $x^h(i,j)$ and $x^v(i,j)$ denote the horizontal and vertical states, respectively, $u(i,j)$ is the input vector, and A_{11}, A_{12}, A_{21}, A_{22}, B_1, and B_2 are real-valued matrices with appropriate dimensions.

The Attasi model is given as [10]

$$
\begin{aligned}
x(i,j) =& A_1 x(i,j-1) + A_2 x(i-1,j) \\
& - A_1 A_2 x(i-1,j-1) + B u(i-1,j-1)
\end{aligned} \qquad (1.2)
$$

with $A_1 A_2 = A_2 A_1$, where $x(i,j)$ is the system state, A_1, A_2, and B are real-valued matrices.

Further, consider the following first Fornasini-Marchesini (FM-I) model [58]:

$$
\vec{x}(i,j) = \vec{A}_1 \vec{x}(i-1,j-1) + \vec{A}_2 \vec{x}(i,j-1) + \vec{A}_3 \vec{x}(i-1,j) + \vec{B} u(i-1,j-1), \qquad (1.3)
$$

where $\vec{x}(i,j)$ is the system state and $\vec{A}_1$, $\vec{A}_2$, $\vec{A}_3$, and $\vec{B}$ are known parameter matrices. In addition, the second Fornasini-Marchesini (FM-II) model is expressed by [57]:

$$
\begin{aligned}
\bar{x}(i,j) =& \bar{A}_1 \bar{x}(i,j-1) + \bar{A}_2 \bar{x}(i-1,j) \\
& + \bar{B}_1 u(i,j-1) + \bar{B}_2 u(i-1,j),
\end{aligned} \qquad (1.4)
$$

where $\bar{x}(i,j)$ is the state vector and $\bar{A}_1$, $\bar{A}_2$, $\bar{B}_1$, and $\bar{B}_2$ are parameter matrices with appropriate dimensions.

1.1.2 Relationships between the Models

Apparently, the Attasi model (1.2) can be derived from the FM-I model (1.3) when $\vec{A}_1 = -A_1 A_2$, $\vec{A}_2 = A_1$, $\vec{A}_3 = A_2$, and $\vec{B} = B$. By defining $x^h(i,j) =$

$\vec{x}(i, j+1) - \vec{A}_2 \vec{x}(i,j)$ and $x^v(i,j) = \vec{x}(i,j)$, model (1.3) is converted into the Roesser model (1.1) with

$$
\left[\begin{array}{cc} A_{11} & A_{12} \\ A_{21} & A_{22} \end{array} \right] = \left[\begin{array}{cc} \vec{A}_3 & \vec{A}_1 + \vec{A}_3 \vec{A}_2 \\ I & \vec{A}_2 \end{array} \right], \quad \left[\begin{array}{c} B_1 \\ B_2 \end{array} \right] = \left[\begin{array}{c} \vec{B} \\ 0 \end{array} \right].
$$

With the aid of some routine manipulations, model (1.1) can be recast into the FM-II model (1.4) with

$$
\bar{A}_1 = \left[\begin{array}{cc} 0 & 0 \\ A_{21} & A_{22} \end{array} \right], \quad \bar{A}_2 = \left[\begin{array}{cc} A_{11} & A_{12} \\ 0 & 0 \end{array} \right]
$$

$$
\bar{x}(i,j) = \left[\begin{array}{c} x^h(i,j) \\ x^v(i,j) \end{array} \right], \quad \bar{B}_1 = \left[\begin{array}{c} 0 \\ B_2 \end{array} \right], \quad \bar{B}_2 = \left[\begin{array}{c} B_1 \\ 0 \end{array} \right].
$$

It is observed that the FM-I model is a special case of the Roesser model, and the FM-II model can be recognized as a more general one which covers the Roesser model. Owing to their broad applications, both the FM-II model and the Roesser model have drawn considerable research interest when dealing with the analysis and synthesis issues of 2-D systems.

1.1.3 Linear Repetitive Processes

As a particular class of 2-D systems, the linear repetitive processes (LRPs) have been gaining momentum owing mainly to their practical insights in industry areas such as machining learning, metal rolling operations, coal mining, and digital allpass filters [13,93,94,144,163,164,205]. A typical repetitive process consists of a series of sweeps (known as passes) which are described by certain differential or difference dynamics over a finite duration (known as the pass length). On each pass, the process output, termed as the pass profile, is developed which performs as a forcing function and contributes to the dynamics of the next pass profile. The distinct feature of an LRP lies in that the state dynamics exhibits along each pass over a finite duration and the pass profile evolves along the pass-to-pass direction. Owing to such a bidirectional transmission, repetitive processes have been greatly investigated with the aid of 2-D theory and some elegant results have appeared [18,161,184,225]. For instance, sufficient criterion has been presented in [161] to ensure that the stability along the pass of the underlying repetitive processes is equivalent to the bounded-input/bounded-output stability of the Roesser model. The observer-based sliding mode control problem has been addressed in [225] for a class of differential LRPs with unknown input disturbance by using the 2-D Lyapunov function.

    The research of repetitive processes is also bound up with iterative control algorithms that pursue to achieve a favorable tracking performance in terms of repetitive operation from circle to circle. There are some classical iterative learning control (ILC) strategies proposed and further applied in

many practical areas, such as target tracking, system guidance, and navigation [17, 89, 92, 134–136, 142, 149, 160]. For example, adaptive ILC schemes
have been designed in [189] for the trajectory tracking problem of rigid robot
manipulators with unknown parameters. The ILC rules have been tackled
in [92] for 2-D systems where convergence of the proposed algorithm has been
discussed. Moreover, an ILC scheme has been considered in [142] for particular
linear systems and the monotone convergence of the tracking error norm has
been investigated based on certain updating control laws. Till now, available
schemes for handling the LRPs could be generally summarized as two cate-
gories. The first one is to develop the 1-D equivalent models with unavoidable lags or augmented dimensions that entirely rely on the pass length [163].
Nevertheless, the massive amount of computation resulting from such a scheme
may be unacceptable from the practical insight. The second one is to entirely
exploit the bidirectional feature by transforming the LRPs under consideration into the 2-D framework [15]. In this regard, the frequency region method
has been introduced to cope with the controllability and observability of the
considered LRPs [52], and 2-D Lyapunov function approach has also been proposed to handle the analysis and synthesis issues of LRPs with linear matrix
inequality characterizations [61].

1.1.4   2-D Models with Other Complicated Dynamics

Fruitful results have been reported on different topics concerning 2-D
systems, while most of the 2-D systems tackled in the literature are supposed to be shift-invariant to simplify the theoretical analysis, where relevant
stability-oriented matters are of major concern. Unfortunately, shift-invariant
models are usually not applicable due to the existence of shift-varying parameters in almost all real-world systems, especially for those plants in heterogeneous environments. Moreover, when taking consideration of the particular
engineering requirements, transient characteristics of systems to be addressed
are often of interest. As such, investigation on 2-D shift-varying systems has
been garnering a growing research interest [151, 196, 261]. In [151], an estimation algorithm has been proposed to estimate the picture element in the displacement field with time-varying imagery by recurring to a spatio-temporal
gradient method, where a number of simplified strategies have been further
developed to reduce the computational complexity of the algorithm, and their
impacts on the estimation performance have also been examined. In [261],
Krein-space-based innovation analysis and projection technique have been exploited to settle the H_∞ fault estimation problem for 2-D linear shift-varying
systems, where existence of the H_∞ fault estimator has been first ensured and
then solution of the estimator has been attained.

On another research front, nonlinearities are ubiquitous in the industrial
domains, and hence nonlinear systems deserve to be analyzed for their promising potentials. The phenomenon of nonlinearities is caused mainly by the
high maneuverabilities of the target plants as well as the frequently occurred

nonlinear disturbances [27]. Involvement of the nonlinearities would lead to identified challenges in addressing the system dynamics. In recent years, some initial research attention has been paid to the 2-D nonlinear systems, and certain preliminary results have been presented, see e.g., [50, 54, 170, 250]. To mention a few, an array of discrete 2-D nonlinear systems described by the Roesser model has been considered in [250], where both asymptotic and exponential stabilities have been discussed based on the Lyapunov theorems. For the positive 2-D Takagi-Sugeno fuzzy systems with state delays, the asymptotic stability and l_1-gain analysis have been addressed in [50] by resorting to the delay-dependent Lyapunov function method.

Enlightened by many available results on complex networks regarding the conventional 1-D models, the 2-D counterparts have been receiving a gradually increasing attention on account of some complex networks exhibiting their dynamics along two directions. To be more specific, a generalized nonorthogonal 2-D transform has been fulfilled in [34] which provides an image description for analysis, segmentation, and compression. A 2-D learning strategy has been proposed in [30] to establish a 2-D networked system for modeling the multilayer neural networks with feed-forward process and learning process in different directions. By means of coupled 2-D chaotic maps on complex networks, the collective dynamics has been studied in [96] for a gene regulatory network of bacterium *Escherichia coli* subjected to different coupling forms and strengths. Moreover, a novel synchronization concept has been introduced in [110] for the 2-D coupled dynamical networks, where sufficient conditions have been presented to ascertain the global synchronization.

1.2 Communication Constraints

Owing to the recent advances on network technologies, more and more filtering/estimation/control algorithms have now been implemented over communication networks with limited capacity under the networked environments [159]. Such network-based implementation has the advantages in simple installation, cost saving, and high flexibility, but it also brings about undesirable network-induced phenomena posing substantial challenges to the system dynamics and performance analysis [86, 209, 238]. There have been many network-induced phenomena including communication delays, packet dropouts, sensor saturations, and signal quantizations [12, 40, 43, 53, 79, 129]. All these phenomena would further challenge the validity of the traditional filtering theory for 2-D systems. Therefore, the filter design problem for 2-D systems with communication constraints has become an intriguing research topic, and numerous efforts have been devoted to investigating the impact of communication constraints on the 2-D filtering issues.

In many real-world scenarios, the communication resources might be limited and costly, and thus it is necessary to mitigate unnecessary waste of these communication energies as much as possible. For effective executions of information transmission, a common prerequisite is that signals from all network nodes can be simultaneously transmitted via the communication channels. Such a prerequisite is, unfortunately, unlikely to be met due to the inevitable data collisions caused by the communication constraints in practice [70, 116]. A rather popular way to prevent data from collisions is to employ the so-called communication protocols to schedule the transmission order of the network nodes. The essence of these scheduling protocols lies in that only part of the nodes is permitted to have the network access at each transmission instant.

1.2.1   Network-Induced Phenomena

Note that traditional filtering theory largely ignores the effects of communication networks or assumes that the communication channels possess sufficient large bandwidth. Nonetheless, it is not uncommon that the signals are sent through wireless channels with limited communication bandwidth in the emerging applications. Consequently, the transmitted signals could be distorted, delayed, lost, and sometimes even not be allowed for transmission. In recent years, an ever-increasing research interest has been focused on the filtering problems for 2-D systems under networked environments, where many kinds of network-induced phenomena have been taken into account and their influences on the estimation performance have been elaborately discussed.

Measurement Degradation

The measurement degradation has been well recognized as a frequently encountered phenomenon resulting typically from the network congestions in the networked environments. The degraded measurements, which include the missing measurements (also called packet losses or dropouts) as a special case, usually occur in a random way because of the random nature of the network load fluctuations, unexpected sensor aging, and accidentally changed working conditions [42, 167, 185, 242]. To depict the missing measurements, the Bernoulli distribution description has been utilized to describe an uncertain observation where the useful information may be missing (containing noises alone) rather than consecutive [145]. Apart from that, a more general way to model the degraded measurement is to introduce a random variable satisfying an arbitrary probabilistic distribution [124].

It is worth mentioning that measurement degradation has been viewed as a key factor in the system dynamics and may seriously degrade the system performance if not appropriately tackled. So far, the optimal H_2 filter has been designed in [166] for networked control systems with multiple packet dropouts based on the linear matrix inequality techniques. The variance-constrained H_∞ filtering problem has been investigated in [44] for nonlinear time-varying systems with missing measurements. In [217], the robust filtering problem has been studied for discrete-time uncertain systems with state-delay and missing measurement, where the missing probability obeys a certain probabilistic distribution over the interval $[0, 1]$. The distributed filtering algorithm has been proposed in [118] for discrete-time systems with stochastic nonlinearities and sensor degradation over a finite horizon, and the mean-square boundedness issue has been further discussed for the estimation errors. With the help of a condition-based maintenance policy, the stochastic filtering has been investigated in [112] to estimate the system state monitored by a degraded sensor. For 2-D systems, the filtering problem with degraded measurements has also drawn some initial attention. For example, the robust state estimation problem has been considered in [109] for 2-D systems with both missing measurement and sensor saturation.

Sensor Delays

Sensor delays are pervasive in the networked systems due mainly to the physical limitations, for instance, the network congestions. Such a phenomenon plays a vital role in studying the system dynamics, which, if not adequately catered for, could degrade the filtering performance and even cause divergence of the filtering algorithm. Up till now, several different communication delays have been introduced in the literature including the constant delay and the time-varying delay [104, 156, 158, 232, 249, 255]. The constant delay is the simplest model used to describe the sensor delay, especially for the case that the buffer in the receiver is longer than the worst-case delay time. Bearing the time-varying property of networked environments into mind, the constant delay may be deficient to depict the sensor delay. The time-varying delay is hence exploited to model the sensor delay in reality by treating it as an uncertainty with known upper and lower bounds. Traditionally, the sensor delay is supposed to occur in a deterministic way. This, however, is not always true. Sensor delays are likely to appear in a random fashion whose probabilistic distribution is determined a priori via statistical tests. The last decade has spotted a great deal of research interest on randomly varying sensor delays; see [111, 172, 212, 262]. The robust filtering problem has been investigated in [212] for a class of stochastic systems with parameter uncertainties and probabilistic sensor delays, and the filters have been designed assuring the mean-square boundedness of the filtering error. The H_∞ filtering problem has been considered in [262] for discrete-time systems with randomly varying sensor delays, where the stochastic stability of the error system and a prescribed H_∞ filtering performance have been guaranteed. Furthermore, the state estimation problem has been addressed in [111] for 2-D complex networks with randomly occurring nonlinearities and sensor delays.

Signal Quantization

Among various network-induced phenomena, signal quantization induced by the bit constraints of the communication channels is one of the frequently encountered behaviors [31]. Signal quantization stemming from the analog-to-digital conversion process is viewed as a contraction mapping. Such a phenomenon is inherently nonlinear and irreversible that ineluctably results in quantization errors acting on the transmitted data. Noting that signal quantization has been recognized to have a great impact on the system dynamics, it makes practical sense to study the quantized systems and analyze the system performance under the quantization effects. Up to now, much work has been done for systems with quantized signals, see e.g., [33, 80, 95, 210, 233, 270].

Roughly speaking, signal quantization can be classified into static quantization and dynamic quantization. Particularly, the static quantization maps the signals into a quantized set by a memoryless nonlinear function. There are generally two ways of modeling the static quantization, namely, the logarithmic quantization (belonging to the floating-point quantization) and the uniform quantization (also known as the fixed-point quantization). The latter has been popular in engineering practice since the extensive utilization of the fixed-point strategy. So far, some elegant results have been reported on the analysis of systems with static quantized measurements [39, 59, 74, 117, 252]. For instance, the coarsest memoryless quantizer has been provided in [74] to stabilize a discrete-time linear system with packet dropouts. The problem of state estimation has been exhaustively analyzed in [59] for linear discrete-time dynamic systems with logarithmic quantizer. In [39], the distributed recursive filtering problem has been investigated for stochastic systems undergoing uniform quantizations and deception attacks over a sensor network with given topology information. The remote state estimation problem has been addressed in [117] for linear discrete time-varying non-Gaussian systems with measurements quantized by a probabilistic uniform quantizer.

Unlike the static quantization, the dynamic quantization generates the quantized signals based on the current and the historical measurements. This fact leads to more complicated dynamics, whereas the dynamic quantization may achieve superior performance over the static one. To date, the dynamic quantization has begun to stir some initial research attention in the context of network-based control or filtering problems [11, 186, 268]. In [11], an optimal dynamic quantizer has been designed in the sense of input–output relation for an array of discrete-valued input systems. The moving horizon estimation problem has been investigated in [268] for networked systems with unknown inputs and dynamic quantization effects, where the ultimate boundedness issue of the estimation error has been presented based on a special decomposition method.

1.2.2 Communication Protocols

In networked systems, the communication channels of limited bandwidth are usually shared by various network components (e.g., sensors, actuators, controllers, and filters). The shared channels make it extremely difficult for information to be simultaneously exchanged/transmitted between the components and accordingly, the data congestions are inevitable that might jeopardize the system performance. An effective way of preventing data congestions, which has been widely adopted in industry, is to introduce the communication protocols to regulate the data transmissions by determining which node has the privilege to access the shared network at each transmission instant [45, 115].

With the advance of network techniques, some representative scheduling schemes for data transmissions have been adopted which mainly embrace the redundant-channel transmissions, the Round-Robin (RR) protocol, the random access (RA) protocol, the try-once-discard (TOD) protocol, and the event-based scheduling mechanism [146, 153, 180, 187, 192, 194]. Generally speaking, the first policy aims at enhancing the network reliability by providing multiple channels for data transmissions, while the others intend to alleviate the communication burden by allowing only part of the measured outputs to be transmitted to the remote controller/filter. It is noteworthy that, according to the way of selecting/triggering network nodes, these policies fall into two distinguishing scheduling rules. More specifically, the first two follow the static scheduling rule (namely, certain predefined transmission orders), whereas the remainders are scheduled dynamically (that is to say, their respective transmission orders are determined in real-time).

Redundant Channel Transmission

As is well known, data packets may be dropped or lost while transmitted over communication networks, which could have negative impacts (e.g., degradation, unreliability, or even instability) on the system performance. To this end, research efforts have been devoted to remedying the undesired effect of the packet dropouts. Particularly, the concept of redundant channels has been proposed to reduce the packet dropout rate, thereby reserving more useful information and improving the system performance [137, 208, 264]. In contrast with data transmitted through a single channel, redundant channels provide two or more available accesses for signal communication. In addition, redundant channels can be rationally ordered according to the packet dropout rates acquired through numerous statistical tests. If one channel is detected to suffer from transmission failures, then the next channel is activated to propagate the received information with purpose of ensuring the communication reliability. Recently, the distributed H_∞ filtering problem has been addressed in [256] for 1-D fuzzy systems with two redundant channels by means of the scaled small gain theorem, where sufficient conditions have been derived to guarantee that the underlying system is stochastically stable and attains a prescribed H_∞

performance index. The measurement model with two channels has been further extended to the multichannel case [29, 41, 203]. In [203], the H_∞ state estimation problem has been solved for multirate time-varying systems with estimation error variance constraints and redundant channels.

RR Protocol

The RR protocol (also referred to as the token ring) appoints equitable privileges for all sensors to transmit signals through the network medium one-by-one in a circular way [46]. Such a protocol is of practical importance since it predefines a periodic transmission rule which makes the scheduling easy-to-implement. In comparison to the conventional communication without protocol, consideration of the RR scheduling would inevitably lead to certain challenges in the analysis/synthesis of the filtering performance. Up to now, some preliminary results have been available for the control/estimation problems with RR scheduling by means of the periodic switching method or accumulated delay approach [114, 127, 265]. In [114], the exponential stability issue has been analyzed for networked control systems with time-varying transmission delays and communication constraints, where the RR protocol has been used to handle the scheduling of sensor information toward the controller. The finite-time distributed state estimation has been studied in [241] for an array of nonlinear systems with RR protocol, where sufficient criteria have been provided to ensure the average stochastic finite-time boundedness and stability for the error system. The distributed recursive filtering issue has been addressed in [174] for stochastic discrete time-varying systems with state saturations and RR protocols over sensor networks.

RA Protocol

Among various communication protocols, the RA protocol has generated great interest from the industry because (1) it belongs to the notable category of carrier-sense multiple access (CSMA) protocols, and (2) it can be readily implemented in various communication systems by using the rule of "sense before transmit". Especially, the CSMA-like protocols have shown their advantages in many practical networks such as WirelessHART and ISA-100 [26, 155], and this motivates the adoption of the RA protocol in real-life applications. It is noteworthy that the RA protocol is applied to the case where at each transmission instant, only one network node (or signal/packet observed by each node) has the privilege of granting access to the shared network (or channel), and which node (or signal/packet) should be selected is based on certain stochastic process. Up to now, some preliminary research results have been reported on the filtering issues of 1-D networked systems with RA protocol, see e.g., [67, 195, 258, 266, 267]. The filter design issue has been studied in [267] for time-varying nonlinear system with stochastic protocol over high-rate communication network, where the randomly switching behavior of the data transmission has been presented and the desired filter has been contrived to meet certain H_∞ disturbance attenuation level. In [266], the recursive filtering problem has been tackled for a class of networked linear time-varying systems suffering from RA scheduling, where the filter gain has been derived by solving two Riccati-like difference equations, and the boundedness issue has also been investigated for the estimation error covariance with lower and upper bounds. For systems with RA protocol scheduling, a rather natural idea is to extend the existing communication-protocol-scheduled filtering algorithms to 2-D systems because of their great application potentials. A thorough literature review has revealed that, unfortunately, the corresponding results have been very few (if not none) on the 2-D recursive filtering problems.

Event-Triggered Mechanism

Concerning communication strategies over networked systems, the traditional time-triggered scheme has been widely utilized for filter design purpose where the communication interval is set to be known *a priori*. From the practical perspective, however, frequent data transmissions would inevitably exhaust the limited bandwidth and further deteriorate quality of the network communication [98, 251]. This is crucial in 2-D systems as the bidirectional transmission nature leads undoubtedly to a large number of information broadcasts. Therefore, the so-called event-triggered schedule has been proposed with hope to reduce the unnecessary system executions, thereby improving efficient usage of the network resources [36, 120, 188]. The core idea of event-triggered transmission is to prorogate meaningful information (only when necessary) at the triggering instants in order to decrease the communication frequency while maintain an acceptable filtering performance. So far, an increasing number of energy-efficient results have been emerging on the filter design issue subject to event-triggered strategies [71, 101, 121, 138, 178, 183, 224, 239]. For example, the send-on-delta data collecting strategy has been employed in [138] to capture the valuable information, where the data sampling strategy is triggered when the signal deviates from a predetermined threshold defined as a significant change. Furthermore, a modified Kalman filter with event-triggered sampling has been proposed in [183] for the networked monitoring plant based on a send-on-delta approach. The stochastic event-triggered sensor schedule has been proposed in [71] to obtain the minimum mean-squared error estimator for guaranteeing the stability of the error covariance. A dynamic event-triggered mechanism has been investigated in [101] to address the recursive distributed filtering problem for discrete nonlinear systems over a sensor network subject to a time-varying topology that is connected via the Gilbert-Elliott channels.

1.3   Recent Progress on Filtering for 2-D Systems

In complex environments such as networked situations, 2-D systems are likely to be contaminated by many kinds of disturbances/noises from different sources. These disturbances/noises can be either deterministic or stochastic with or without precisely known statistics. A notable type of deterministic disturbances is the energy-bounded disturbance caused mainly by unexpected environmental changes or external interferences. The stochastic noises are also frequently involved in the underlying systems, for example, the Gaussian white noises. It is worth mentioning that the stochastic terms are not uncommon to be found in the considered systems since the stochastic phenomena are unavoidable in reality. Roughly speaking, the stochastic phenomena could be (1) random noises or perturbations stemming from abrupt variations of the internal system components and/or sudden changes of the external working conditions and (2) randomly occurring information including time delay, packet dropout, sensor failure, channel fading, and signal quantization that are encountered in a random manner.

Estimating the system states that cannot be observed directly is a long-term project in the control theory and signal processing [130]. Concerning 2-D systems, their filtering problems have been proven to be an appealing and active research topic, whose focus is to estimate/reconstruct the internal states of interest based on the obtainable but corrupted measurements [77, 207]. As such, the filtering problem for various types of 2-D systems has recently become a hot research topic that attracts a recurring interest, and several filtering methods are principally adopted to evaluate the estimation accuracy.

1.3.1 H_∞ Filtering

The H_∞ filtering scheme is a favored choice for systems with energy-bounded disturbances [8, 173]. The intention of H_∞ filtering is to cater for a specified H_∞ index that bounds the disturbance attenuation level (namely, the L_2 gain from the external inputs to the estimation errors) in the worst case. So far, a large body of literature has been published on the H_∞ filtering problems for many 2-D systems based on the linear matrix inequality techniques, see e.g., [72, 102, 107, 259].

In the pioneering work [72], a sufficient condition in terms of the Lyapunov equation has been first presented for ensuring the asymptotic stability of the 2-D shift-invariant system in FM-II model, and then the established stability condition has been exploited to design the 2-D digital filter. Subsequently, the results obtained in [72] have been extended in [125] by developing a generalized Lyapunov equation, and an easy-to-be satisfied condition has been provided for designing the 2-D filter with guaranteed asymptotic stability. In [49], the H_∞ filtering problem has been dealt with for 2-D systems in the form of FM-II model with invariant parameters and energy-bounded inputs. Particularly, a 2-D version of the bounded real lemma has been found in [49] which is competent for solving the finite horizon and infinite horizon filtering problems. Further based on the 2-D bounded realness property, the H_∞ deconvolution filtering problem has been studied in [235] for a class of 2-D digital systems, and the design of 2-D deconvolution filter has been casted into a convex optimization problem in terms of linear matrix inequalities. For 2-D shift-invariant systems in the Roesser model with polynomial uncertainties, the related H_2 and H_∞ norms have been introduced in [191], and the robust mixed H_2/H_∞ filtering problem has been solved by the convex linear matrix inequality method.

Note that the system parameters tend to be parameter-varying on account of the structural changes in models. Typical examples include repairs/failures of intermittent components, and changes/interruptions of internal subsystems. A popular way for modeling the parameter-varying systems relies on the unknown but measurable parameters, whose variations are given in the prescribed intervals. Besides, Markovian jump models have also been broadly applied to describe the parameter-varying systems. In this case, the gain-scheduled method has been widely adopted to construct the parameter-dependent filters. Generally, the Markovian jump model is an array of subsystems with certain transitions between the subsystems, where the transitions are governed by a Markov chain, and thus the system matrices randomly change and take values in a finite set. The H_∞ filtering issue has been considered in [227] for 2-D Markovian jump systems described by the Roesser model with fully available transition probabilities, where sufficient conditions have been presented assuring the filtering error system to be asymptotically stable with a certain H_∞ performance specification. In [218], the 2-D state-delay systems with deficient modes in the Markov process have been discussed, in which an H_∞ performance criterion has been first developed and then the H_∞ filter has been constructed by elaborately using the transition probability matrix.

Some 2-D systems in engineering applications are inherently nonlinear. The filtering problems related to 2-D nonlinear systems are generally challenging due to the complicated dynamics caused by nonlinearities. Accordingly, research on the filter design of 2-D nonlinear systems has become an attractive topic from both the theoretical and the engineering viewpoints. The literature [113] has been concerned with the design of 2-D digital filter with overflow nonlinearities by the Lyapunov method. The fuzzy logic theory as an efficient approach handling the nonlinear systems has also been utilized in 2-D case. For instance, the robust H_∞ filtering problem has been considered in [126] for 2-D fuzzy systems with fractional-type uncertainties and randomly occurring mixed delays, where the estimation error has been verified to be asymptotically stable with a prescribed H_∞ index under certain conditions, and the filter parameters have also been given by the solution to a particular convex optimization problem. Besides, design of the H_∞ filters has been discussed for

2-D fuzzy systems in [16, 99] with aid of the fuzzy theory and Lyapunov-like function.

1.3.2 l_2-l_∞ Filtering

The l_2-l_∞ filtering is another preferred candidate for tackling the energy-bounded inputs. Different from the H_∞ filtering which aims at the energy-to-energy performance, the l_2-l_∞ scheme is concerned with the achievement of energy-to-peak performance [23]. Particularly, the l_2-l_∞ filtering concentrates on the design of an energy-to-peak gain in terms of suppressing the amplitude of the variations concerning the system outputs or optimizing the peak value of the estimation errors. The l_2-l_∞ filtering is also referred to as a generalized H_2 approach or a deterministic procedure of the Kalman filtering [165]. To cope with the energy-to-peak filtering problem, the l_2-l_∞ scheme has been first introduced in [150] for the 1-D linear systems with convex-bounded uncertainties and then applied in [64] for the 1-D uncertain systems with time-varying state delays. Afterwards, a full-order l_2-l_∞ filter has been designed in [230] for the 2-D linear parameter-varying system, where the asymptotic stability of the error dynamics and the prescribed disturbance attenuation level have been gained by solving a certain optimization problem. In [2], the l_2-l_∞ stability criterion has been provided for the 2-D digital filters with external interferences. By means of the l_2-l_∞ method, the non-fragile fault estimation problem has been investigated in [128] for the 2-D Markovian jumping systems, where the filtering performances with respect to the energy-to-peak gain and the prescribed power bound constraints have been carefully discussed.

1.3.3 l_1 Filtering

Unlike the energy-bounded noises considered in the H_∞ and l_2-l_∞ filtering settings, the disturbance signals can be peak-bounded, and an effective estimation strategy for handling such kind of noises is the l_1 filtering for discrete-time systems (or the L_1 filtering for continuous-time systems). The l_1 filtering focuses on the peak-to-peak gain so as to minimize the worst-case peak value of the estimation error for systems with peak-bounded disturbances. In other words, the objective of the l_1 filtering lies in the achievement of the peak-to-peak performance. The earliest result on the peak-to-peak performance has been presented in [193] to tackle the optimal rejection problem with persistent bounded disturbances, where the maximum amplitude of the system output has been minimized. Inspired by the pioneering works in [1], the l_1 filtering issues have attracted a recurring interest in recent years. Particularly, a peak-to-peak filter has been designed in [24] for networked nonlinear DC motor systems, which ensures the asymptotic stability of the error dynamics and the realization of the specified performance index. Despite these appealing results for 1-D systems, the peak-to-peak filtering strategy has also been proposed in [3] for the 2-D stochastic systems subject to multiplicative noises, where sufficient conditions have been derived guaranteeing the mean-square stability of the underlying system as well as the peak-to-peak performance.

1.3.4 Dissipative Filtering

The dissipative systems were initially introduced by Willems in the 1970s [220, 221]. By introducing the quadratic energy supply function, the dissipativity theory reflects the dissipativity of the system with an input-output energy-related description. In this sense, the dissipativity theory supplies a unified framework for analyzing the controller and filter design issues of systems with interferences. It is worth noting that the dissipative filtering is an extension of the H_∞ filtering scheme, and the dissipativity analysis covers the passivity problem. To date, the system dissipativity and its applications have been widely investigated in the area of signal processing, and many basic yet important results have been reported. The problem of α-dissipativity analysis has been studied in [56] for time-delay singular systems by means of the delay partitioning technique. Then, a generalized dissipativity concept has been developed in [253] for 1-D continuous Markovian jump systems, where the extended dissipativity has been provided as a new performance index that incorporates the H_∞, L_2-L_∞, passive and dissipative filtering performances in a unified form.

The dissipative filtering has also acquired some initial consideration for 2-D systems. In [4], the concept of 2-D dissipativity has been defined for a linear discrete-time 2-D system in the form of Roesser model, and the 2-D dissipative control and filtering problems have been solved based on the proposed 2-D dissipativity notion. In that paper [4], both the 2-D passivity and the H_∞ performance have been discussed as special cases of the defined 2-D dissipativity. Later on, the generalized dissipativity of digital filters has been explored for a class of FM-II models with saturation functions, and the stability issue of the unforced 2-D filters has been further examined in [5]. Furthermore, in [97], the dissipative filtering problem has been considered for 2-D linear systems with norm-bounded uncertainties, where a non-fragile dissipative filter has been constructed with achievements of a strict dissipative performance index and the asymptotic stability for the addressed model.

1.3.5 Kalman Filtering

The celebrated Kalman filter has a recursive fashion and is famous for its capacity of achieving the minimum variance of the estimation error at each iteration. Originated from its invention in [85], the Kalman filtering algorithm has been well studied for a variety of 1-D systems with extensive applications ranging from power systems, tracking systems, navigation to spacecraft and satellite [25, 75, 139, 143]. Inspired by this truth, an extension of the Kalman filtering from the 1-D framework to the 2-D case has been a concern. One of the first attempts has been made in [223], where the Kalman filtering has been investigated for image processing over a discrete 2-D square region. Two approximate Kalman filtering schemes have also been introduced in [223] to reduce the excessive computational load, where the first filtering scheme viewed as the strip processor updates a line segment at each step, and the other (also termed as the reduced update Kalman filter) is a scalar processor updating only the nearby states of those latest processed states. Soon after, the reduced update Kalman filtering method has been further used to solve the deconvolution problem of image restoration in [222]. A new approximate 2-D recursive filtering has been developed in [87] for a stationary random field modeled by a 2-D system. Moreover, the recursive filtering problem has been handled in [32] for noisy nonhomogeneous images, where a semi-causal filter has been proposed by using the edge preserving properties.

Despite of these representative works, the proposed 2-D filtering algorithms suffer from restrictive conditions and are computationally wasteful or even unmanageable. It should be pointed out that the 2-D Kalman filtering algorithms in the mentioned literature can be recognized as a combination of the sequences of 1-D fixed-interval smoothers and the 1-D Kalman predictors, which cannot provide a systematic framework of the 2-D Kalman filters. To better reflect the two-directional characteristic, the Kalman filtering algorithm with a modest computational burden has been developed in the seminal literature [269], where a 2-D Kalman filter minimizing the estimation error variance has been successfully designed for the 2-D shift-invariant system with additive noises. The Kalman filtering problem without any structure constraint on the filter structure has been addressed in [157] for a class of 2-D linear systems. In addition, the minimum-variance filtering problem has been considered in [105] for the 2-D shift-varying systems subjected to stochastic non-linearities and degraded measurements, where the filter parameters have been subtly determined and the estimation performances regarding the boundedness and monotonicity have been further analyzed.

Note that the classical Kalman filter is applied to linear systems with Gaussian noises, where both the system parameters and the statistics of the Gaussian noises are assumed to be known accurately. For systems violating either the linear structure or the exact knowledge of the parameter matrices and the noise statistics, the Kalman filtering strategy becomes inapplicable because it probably results in a debased estimation performance. To complement the traditional Kalman filter and widen its utilization ranges, some efficient methods have been introduced in the literature and gained a firmly growing concern in the fields of control theory and engineering applications [28, 76, 82, 169].

1.3.6 Variance-Constrained Filtering

For the underlying systems, parameter uncertainties or variations may appear because of the unmodeled system dynamics. As pointed out in [9], even slight modeling uncertainties/errors, if not carefully addressed, could seriously worsen the estimation performance. Furthermore, in certain cases where the system model undergoes incomplete information (e.g., parameter uncertainties or imperfect measurements), it might be theoretically difficult (also practically unnecessary) to acquire the global optimality (minimum-variance) of the state estimation [211, 215]. An alternative way is then to obtain a suboptimal performance by ensuring a tight upper bound on the estimation error variance. One of the typical examples is the target tracking problem where the upper bound corresponds to the size of the tracking window, see [213] for more details.

With the aim of enhancing the robustness of the standard Kalman filtering, the variance-constrained filtering has been garnering much research attention in the past two decades [182, 190, 214, 243]. Different from the Kalman filtering which ensures the minimum error variance for systems with Gaussian white noises, the basic concept of variance-constrained filtering is to derive an upper bound on the error variance and then design proper gain parameters rendering this bound to be optimal. To date, much research attention has been paid to the filtering problems with guaranteed (minimized) or constrained (prescribed) upper bounds on the error variances [60, 141, 237, 244]. For instance, the design problem of robust Kalman filter has been coped with in [237] for an array of discrete-time systems with time-varying norm-bounded parameter uncertainties in both the state and the output matrices. The finite-horizon robust Kalman filtering problem has been carried out in [60] for uncertain systems, where the recursive filter has been proposed and the optimal scaling parameters can be computed by solving a semi-definite program. A variance-constrained filtering issue has been coped with in [244] for time-varying uncertain systems with both additive and multiplicative noises, where the recursive filter has been designed for achieving an optimized bound on the error variance in terms of two Riccati difference equations.

Almost all available results of variance-constrained filtering focus on 1-D systems, the 2-D counterpart has received scattered research attention and only a few results have been made at its initial stage. The robust Kalman filtering problem has been tackled in [106] for 2-D shift-varying uncertain systems with stochastic noises. After that, a robust finite-horizon filtering method has been investigated in [198] for a class of 2-D shift-varying systems with simultaneous presence of norm-bounded parameter uncertainties and randomly occurring sensor delays. In [260], the robust Kalman filtering problem has been studied for 2-D uncertain systems, where the original problem has been turned into a deterministic robust regularized problem by the least squares method.

For stochastic systems involving nonlinearities, a renowned variant based on Kalman filter has been set up as the extended Kalman filter, which has been well-established for the 1-D nonlinear systems. Then, a natural idea arisen is to develop an extended Kalman filtering for the 2-D nonlinear systems. Recently, the recursive filtering problem has been considered in [197] for the 2-D nonlinear systems suffering from degraded measurements over a finite horizon, where the minimal bound on the error variance has been attained with desired filter gains and influence of the measurement degradations on the filtering performance has been further elaborated.

1.3.7   Protocol-Based Filtering

When addressing the filtering issues of networked systems, it should be noticed that the sensors suffer from restricted sensing/communication capacities because of the limitations of network resources, then simultaneous data transmissions would inevitably lead to network congestions. Moreover, execution of the information interaction at every instant tends to bring out redundant/ useless data and even exhausts the energy cost. To efficiently transmit signals, a practical way is to convey only part of the signals observed from the sensors rather than simultaneously release all of them to the shared network. Therefore, the protocol-based filtering stands out as a delightful choice for its superiority of communication scheduling. Compared with other filtering schemes, the protocol-based filtering aims to obtain an acceptable filtering performance meanwhile mitigate the communication burden and save the network resources.

Nowadays, a variety of communication protocols [122, 204] and energy-efficient mechanisms [37, 119] have been extensively resorted to when the protocol-based filtering problems are of concern, which have also stimulated the design of protocol-based filters for 2-D systems. Among them, the transmission scheme with redundant channels is of significance to reduce the occurrence of packet dropouts. Recently, the redundant-channel transmission has been considered in [199] for 2-D state estimation problem on the basis of mathematical induction and variance-constrained approach. As another useful strategy, the RR protocol scheduling allocates equivalent privilege for each node in a periodic manner. For its implement simplicity, the RR scheduling has been introduced in [127] for 2-D uncertain nonlinear systems, and the robust H_∞ fault detection problem has been investigated. Afterwards, new periodic scheduling has been proposed in [201] for LRPs subjected to uniform quantization and random gain variations. Unlike the redundant channels and the RR protocol served as static scheduling, the RA protocol and the TOD protocol orchestrate the transmission order dynamically. In [202], the RA protocol randomly scheduling the transmission traffic of the shared network has been applied to handle the distributed filtering problem for 2-D shift-varying systems. For the TOD scheduling, only one of the sensors is permitted to transmit its information, and a quadratic selection principle is adopted to identify which sensor acquires the access at each transmission step. Very recently, a multirate LRP with TOD protocol has been discussed in [175], where a fusion estimator has been gained on the strength of the lifting technique and the sequential covariance intersection fusion approach.

Apart from that, the event-triggered mechanism is effective in decreasing the frequency of unnecessary data transmissions. As an aperiodic scheme, the event-triggered mechanism executes the data transfer exclusively when the deviation between the current and the latest transmitted data satisfies a predetermined condition. In these days, the event-triggered filtering problems have been considered for various systems with communication constraints, especially for the traditional 1-D systems [38]. By contrast, the relevant outcomes have been scattered for the energy-efficient filter design of 2-D systems. Literature [200] has made one of the first attempts to discuss the recursive filtering problem for repetitive processes with an event-based mechanism. A novel triggering-shift sequence and a triggering law have been adopted in [200] for the target plant under which the event-based filter gain has been precisely attained.

1.4 Outline

The outline of this book is given as follows. It is worth mentioning that Chapter 2 focuses on the minimum-variance recursive filtering problem for 2-D shift-varying systems. Chapters 3–5 are concerned with the robust Kalman filtering problems for 2-D systems subjected to parameter uncertainties. Chapter 6 investigates the 2-D resilient filtering problem with redundant channels, while Chapter 7 studies the 2-D distributed filtering problem over sensor networks. Chapters 8 and 9 discuss the recursive filtering problems for linear shift-varying repetitive processes. The framework of this book is shown as follows:

- Chapter 1 firstly introduces the research background of 2-D systems, secondly presents some concepts concerning communication constraints, then reviews research progresses and challenges of the filtering problems for 2-D systems with communication constraints, and finally lists the outline of this book.

- Chapter 2 addresses the recursive minimum-variance filtering problem for a class of 2-D shift-varying systems with stochastic nonlinearity and degraded measurements. Utilizing an inductive approach, unbiasedness of the proposed filter is firstly ensured, and parameters of the filter are then designed by resorting to the completing-the-square method. Subsequently, the filtering performances including the boundedness and monotonicity are investigated with respect to the measurement degradations through mathematically rigorous analysis. Moreover, a computational algorithm is presented to facilitate the online implementation of the designed filter.

- Chapter 3 investigates the robust Kalman filtering problem for a class of 2-D shift-varying uncertain systems with both additive and multiplicative noises. Recursion of the generalized estimation error variances for the addressed 2-D system is first established by introducing a 2-D identity quadratic filter, based on which an upper bound of the generalized estimation error variance is obtained. Subsequently, such a bound is minimized in the trace sense by properly designing the filter parameters. The design scheme of the robust Kalman filter is presented in terms of two Riccati-like difference equations that can be recursively computed for programmed applications.

- Chapter 4 deals with the robust finite-horizon filter design problem for 2-D shift-varying systems with norm-bounded parameter uncertainties and incomplete measurements. With the aid of the inductive approach and the 2-D Riccati-like difference equations, sufficient conditions are provided ensuring the existence of an upper bound on the estimation error variance, an algorithm is then developed to derive such an upper bound, and finally the desired filter is designed to minimize the obtained upper bound. The filter design procedure is in a recursive form that facilitates the online calculation.

- In Chapter 5, the recursive filtering problem is handled for a class of 2-D systems suffering from missing measurements with uncertain probabilities. The dynamics of the error variances are presented firstly. Then, by utilizing the stochastic analysis and inductive method, an upper bound is established for the filtering error variance and, subsequently, the minimal one is achieved at each step by choosing a suitable filter gain. The desired upper bound can be obtained recursively by solving two sets of difference equations.

- Chapter 6 investigates the state estimation problem for 2-D shift-varying systems with redundant channels, where the implemented estimator gain is subject to stochastic perturbations. By employing the induction method and the variance-constrained approach, an upper bound on the estimation error variance is firstly constructed and, subsequently, a locally minimal upper bound is achieved by appropriately designing the gain parameter.

- Chapter 7 is concerned with the distributed filtering problem for 2-D shift-varying systems over sensor networks subject to RA protocol. Recursive distributed filters are proposed to estimate the system state through available information from both the individual and the neighboring nodes. Sufficient conditions are first established on the existence of an upper bound of the error variance. Then, by means of a matrix simplification technique, the desired filter gains are designed to optimize the obtained upper bound at each shift step.

- In Chapter 8, the recursive filtering problem is investigated for a class of linear shift-varying repetitive processes with communication constraints. The RR protocol scheduling is applied to orchestrate the transmission order of sensor nodes in a periodic manner. The repetitive processes are first cast into a general FM-II model by using the lifting technique. Sufficient condition is then provided to ensure the local minimization of certain upper bound on the filtering error variance. Furthermore, the boundedness issue is discussed with respect to the filtering error variance.

- In Chapter 9, the event-triggered filtering problem is studied for linear shift-varying repetitive processes with limited network resources. A new definition of the triggering-time sequence is introduced and an event-triggered rule is then constructed for the transformed system. With the aid of mathematical induction, the estimation error variance is guaranteed to have an upper bound which is then minimized with appropriate filter parameters. Theoretical analysis further reveals the monotonicity of the filtering performance with regard to the event-triggered threshold.

- In Chapter 10, conclusions and some potential research directions for future work are given.

2

Minimum-Variance Recursive Filtering for 2-D Systems with Degraded Measurements: Boundedness and Monotonicity

Benefiting from the inherent feature of bidirectional propagations, 2-D systems have shown their extensive application potentials in modeling certain real-world systems exhibiting dynamic behaviors on multiple independent variables. To date, the filtering problem for 2-D systems has gained considerable research interest and several filtering techniques have been proposed. As one of the most popular estimation approaches, a typical Kalman filtering aims at constructing an optimal filter for 1-D linear systems with guaranteed minimum variance of the estimation error in the form of recursive Riccati equations facilitating online calculation. The Kalman filtering has been well investigated for many 1-D systems with successful applications, and a seemingly natural idea is thus to extend the Kalman filtering algorithm to the 2-D counterparts.

Note that nonlinearities are ubiquitous phenomena in engineering practice which give rise to complexity in system analysis and synthesis. The concept of the so-called stochastic nonlinearities has been proposed in [81, 248], where the second-order statistics are given to depict the occurrence possibility of nonlinearities. In addition to the nonlinearities, degraded measurements serve as another important factor that complicates the design of filters/estimators and contributes to undesired filtering performance if not properly handled. The recursive filter design problem has not been fully considered for 2-D shift-varying systems yet, let alone the case when both stochastic nonlinearity and degraded measurements become the major concern.

Motivated by the above discussions, this chapter investigates the recursive filtering problem for 2-D shift-varying system with stochastic nonlinearity and degraded measurements over a finite horizon, where the stochastic nonlinearity is statistically characterized and the degraded measurements are described by random variables with certain probabilistic distributions. Bearing in mind that the dynamics of the 2-D systems propagates along two independent directions, three research questions arise naturally as follows: (1) how to design a suitable filter with optimal gain parameters so as to ensure the minimum filtering error variance at each shifting step? (2) how to evaluate the influence of the measurement degradation on the filtering performance? (3) how to establish an effective algorithm to implement the developed recursive filter? It is, therefore, the main aim of this chapter to provide satisfactory answers to the questions raised above.

In this chapter, for the addressed 2-D system, an unbiased recursive filter is first developed and the optimal filter parameters are then determined to minimize the estimation error variance at each step by utilizing the completing-the-square method. Theoretical analysis results are also provided for the filtering performance. The main contributions of this chapter are highlighted as follows: (1) an unbiased minimum-variance filter is designed for the 2-D shift-varying system subject to stochastic nonlinearity and degraded measurements; (2) both boundedness and monotonicity of the filtering performance are analyzed with rigorous mathematical deduction; and (3) an effective on-line algorithm is presented to realize the proposed recursive filtering which can be iteratively computed.

The remainder of this chapter is outlined as follows. In Section 2.1, the considered 2-D system is introduced and the relating filtering problem is formulated. Section 2.2 presents our main results for the optimal filter design. In Section 2.3, the filtering performance is analyzed which includes both the boundedness and the monotonicity, and the corresponding filter implementation algorithm is also provided. A numerical example with practical background is given in Section 2.4 to demonstrate the effectiveness of the designed filter. Conclusions are finally drawn in Section 2.5.

2.1   Problem Formulation

Consider a 2-D stochastic system described by the following general FM-II model:

$$
\begin{aligned}
x(i,j) =\ & A_1(i,j-1)x(i,j-1) + A_2(i-1,j)x(i-1,j) \\
& + f(x(i,j-1),\xi(i,j-1)) + f(x(i-1,j),\xi(i-1,j)) \\
& + B_1(i,j-1)w(i,j-1) + B_2(i-1,j)w(i-1,j), \qquad (2.1a) \\
y(i,j) =\ & \Xi(i,j)C(i,j)x(i,j) + v(i,j) \qquad\qquad\qquad\qquad\quad (2.1b)
\end{aligned}
$$

where, for $i,\ j \in [1\ N]$ with N being a given positive integer, $x(i,j) \in \mathbb{R}^n$ is the state vector and $y(i,j) \in \mathbb{R}^m$ is the measurement output, and $w(i,j) \in \mathbb{R}^p$ and $v(i,j) \in \mathbb{R}^m$ are zero mean Gaussian white-noise sequences with covariance matrices $Q(i,j) \geq 0$ and $R(i,j) > 0$, respectively. $A_l(i,j)$, $B_l(i,j)$ $(l = 1,2)$, and $C(i,j)$ are known shift-varying matrices, whereas $\Xi(i,j) \triangleq \mathrm{diag}\{\beta_1(i,j),\beta_2(i,j),\dots,\beta_m(i,j)\}$ with $\beta_r(i,j)$ $(r \in [1\ m])$ being random variables that take values on $[0,1]$ with known statistical properties $\mathbb{E}\{\beta_r(i,j)\} = \bar{\beta}_r(i,j) > 0$ and $\mathrm{Var}\{\beta_r(i,j)\} = \tilde{\beta}_r(i,j)$. Here, $\bar{\beta}_r(i,j) > 0$ is referred to as the degradation probability.

The nonlinear function $f(x(i,j), \xi(i,j))$ possesses the following properties:

$$f(0, \xi(i,j)) = 0, \tag{2.2a}$$

$$\mathbb{E}\{f(x(k,l), \xi(k,l)) | x(i,j)\} = 0,$$
$$(k,l) \in \{(k_1, l_1) | k_1 > i \text{ or } l_1 > j\} \cup (i,j) \tag{2.2b}$$

$$\mathbb{E}\{f(x(i,j), \xi(i,j)) f^T(x(k,l), \xi(k,l)) | x(i,j)\}$$
$$= \sum_{s=1}^{h} \Pi_s x^T(i,j) \Gamma_s x(i,j) \delta(i,k) \delta(j,l), \tag{2.2c}$$

where $\xi(i,j) \in \mathbb{R}^{n_\xi}$ with n_ξ being a given positive integer is a zero mean random sequence with variance $\sigma^2 I$, $h > 0$ is a given integer, Π_s, Γ_s ($s \in [1 \ h]$) are known matrices, and $\delta(i,k)$ is the Kronecker delta function with $\delta(i,k)$ being unity for $i = k$ but zero elsewhere.

Set $\eta_r(i,j) \triangleq [w^T(i,j) \ v^T(i,j) \ \beta_r^T(i,j) - \bar{\beta}_r^T(i,j) \ \xi^T(i,j)]^T$. For simplicity, it is assumed that, for $i,k,j,l \in [0 \ N]$,

$$\mathbb{E}\{\eta_r(i,j) \eta_t^T(k,l)\}$$
$$= \text{diag}\{Q(i,j), R(i,j), \tilde{\beta}_r(i,j) \delta(r,t), \sigma^2 I\} \delta(i,k) \delta(j,l). \tag{2.3}$$

The initial boundary conditions for system (2.1), which are independent of all the stochastic variables $\xi(i,j)$, $\beta_r(i,j)$ ($r \in [1 \ m]$), $w(i,j)$, and $v(i,j)$ for all $i,j \in [0 \ N]$, satisfy the following statistical constraints for $i,k,j,l \in [0 \ N]$:

$$\mathbb{E}\{x(i,0)\} = u_1(i), \quad \mathbb{E}\{x(0,j)\} = u_2(j), \tag{2.4a}$$

$$\text{Cov}\{x(i,0), x(k,0)\} = P(i,0)\delta(i,k), \tag{2.4b}$$

$$\text{Cov}\{x(0,j), x(0,l)\} = P(0,j)\delta(j,l), \tag{2.4c}$$

$$\text{Cov}\{x(i,0), x(0,j)\} = P(0,0)\delta(i,0)\delta(0,j), \tag{2.4d}$$

where $u_1(i)$, $u_2(j)$, $P(i,0)$, and $P(0,j)$ are known parameters with appropriate dimensions and $u_1(0) = u_2(0)$.

Remark 2.1 *The stochastic nonlinearity $f(x(i,j), \xi(i,j))$, which is a 2-D version of the one proposed in [81,248], is general enough to encompass many special cases such as the state-dependent multiplicative noises and random variables with their power relying on the sign of the state. On the other hand, an incomplete observation $y(i,j)$ is considered in this chapter in order to account for the possible occurrence of the measurement degradation. Specifically, by introducing the random matrix $\Xi(i,j)$, the measurement degradation of the r-th element of measurement $y(i,j)$ is signified by the probabilistic coefficient $\beta_r(i,j)$ taking values on $[0,1]$ with known statistical properties. Note that, if $\beta_r(i,j)$ takes values on either 0 or 1, then the degraded measurements reduce to the missing ones modeled by the Bernoulli distributed white sequence proposed described as early as in [145].*

Remark 2.2 *The shift-varying system (2.1) embracing stochastic nonlinearity and degraded measurement provides a general FM-II structure with the goal of depicting certain practical systems defined on multiple horizons. Particularly, many industrial processes including the image processing, heating process and fluid dynamics can be described by specific partial differential equations (PDEs). By sampling the functions in these differential equations with respect to certain spatial coordinates in an appropriate way, the PDEs under consideration can be approximately converted into the FM-II model. Attention should also be paid to the fact that the target plant might be affected by undesired working conditions such as fouling, dust and irregular fluctuation. Moreover, observations probably undergo degradations in view of the sensor failures. In this setting, the considered system (2.1) definitely has a promising applicability in practice.*

In this chapter, a two-step recursive filter is adopted as follows:

$$\hat{x}_p(i,j) = A_1(i,j-1)\hat{x}_u(i,j-1) + A_2(i-1,j)\hat{x}_u(i-1,j), \qquad (2.5a)$$

$$\hat{x}_u(i,j) = \hat{x}_p(i,j) + K(i,j)\left[y(i,j) - \bar{\Xi}(i,j)C(i,j)\hat{x}_p(i,j)\right], \qquad (2.5b)$$

where $i,j \in [1\ N]$, $\hat{x}_p(i,j)$ is the one-step prediction of state $x(i,j)$, $\hat{x}_u(i,j)$ is the corresponding updated estimate with initial conditions $\hat{x}_u(i,0) = u_1(i)$ and $\hat{x}_u(0,j) = u_2(j)$ for $i,j \in [0\ N]$, matrix $\bar{\Xi}(i,j) \triangleq \mathbb{E}\{\Xi(i,j)\}$, and $K(i,j)$ is the filter parameter to be designed.

Remark 2.3 *Note that the innovation information $y(i,j) - \bar{\Xi}(i,j)C(i,j)\hat{x}_p(i,j)$ is utilized to update the estimate as shown in (2.5b). The recursive filter proposed here is also referred to as a Kalman-type filter since it has a similar structure to the classical Kalman filter for the 1-D systems.*

Denote the prediction error as $\tilde{x}_p(i,j) \triangleq x(i,j) - \hat{x}_p(i,j)$ and the filtering error as $\tilde{x}_u(i,j) \triangleq x(i,j) - \hat{x}_u(i,j)$. Then, it follows from (2.1), (2.5a), and (2.5b) that

$$\begin{aligned}
\tilde{x}_p(i,j) &= A_1(i,j-1)\tilde{x}_u(i,j-1) + A_2(i-1,j)\tilde{x}_u(i-1,j) \\
&\quad + f(x(i,j-1),\xi(i,j-1)) + f(x(i-1,j),\xi(i-1,j)) \\
&\quad + B_1(i,j-1)w(i,j-1) + B_2(i-1,j)w(i-1,j), \qquad (2.6a)
\end{aligned}$$

$$\begin{aligned}
\tilde{x}_u(i,j) &= (I - K(i,j)\bar{\Xi}(i,j)C(i,j))\tilde{x}_p(i,j) - K(i,j) \\
&\quad \times \left[(\Xi(i,j) - \bar{\Xi}(i,j))C(i,j)x(i,j) + v(i,j)\right]. \qquad (2.6b)
\end{aligned}$$

The objective of this chapter is to design the recursive filter (2.5) such that the filtering error variance $\mathbb{E}\{\tilde{x}_u(i,j)\tilde{x}_u^T(i,j)\}$ is minimized in the sense of spectral norm at each shifting step (i,j) for $i,j \in [0\ N]$. This problem is referred to as the finite-horizon recursive filtering problem for the 2-D stochastic system (2.1).

2.2   The Minimum-Variance Filter Design

In this section, we aim to solve the recursive filtering problem for the addressed 2-D system (2.1). The optimal filter is developed to achieve the minimum error variance by selecting appropriate gain parameters.

Before presenting the main results, we first deal with the unbiasedness of the filter as well as the recursions of the second-order moment for state and the prediction error variances. For notational simplicity, we define

$$X(i, j) \triangleq \mathbb{E}\{x(i, j)x^T(i, j)\},$$
$$P_p(i, j) \triangleq \mathbb{E}\{\tilde{x}_p(i, j)\tilde{x}_p^T(i, j)\},$$
$$P_u(i, j) \triangleq \mathbb{E}\{\tilde{x}_u(i, j)\tilde{x}_u^T(i, j)\}.$$

To start with, the following lemma is introduced that will be used in deriving our main results.

Lemma 2.1 *[73] Let $A = (a_{ij})_{n \times n}$ be a real-valued matrix and $B = \text{diag}\{b_1, b_2, \ldots, b_n\}$ be a diagonal random matrix. Then, we have*

$$\mathbb{E}\{B^T AB\} = \begin{bmatrix} \mathbb{E}\{b_1^2\} & \mathbb{E}\{b_1 b_2\} & \cdots & \mathbb{E}\{b_1 b_n\} \\ \mathbb{E}\{b_2 b_1\} & \mathbb{E}\{b_2^2\} & \cdots & \mathbb{E}\{b_2 b_n\} \\ \vdots & \vdots & \ddots & \vdots \\ \mathbb{E}\{b_n b_1\} & \mathbb{E}\{b_n b_2\} & \cdots & \mathbb{E}\{b_n^2\} \end{bmatrix} \circ A.$$

The following lemma ensures the unbiasedness of filter (2.5) to facilitate the analysis given later.

Lemma 2.2 *For the 2-D system (2.1), the proposed filter in the form (2.5) is unbiased, i.e., $\mathbb{E}\{\tilde{x}_u(i, j)\} = 0$ for $(i, j) \in \mathscr{S}[0\ N]$ and $\mathbb{E}\{\tilde{x}_p(i, j)\} = 0$ for $(i, j) \in \mathscr{S}[1\ N]$.*

Proof *The proof is carried out by mathematical induction. In view of the initial values $\hat{x}_u(i, 0) = u_1(i) = \mathbb{E}\{x(i, 0)\}$ and $\hat{x}_u(0, j) = u_2(j) = \mathbb{E}\{x(0, j)\}$, it is easy to see that, for $i, j \in [0\ N]$,*

$$\mathbb{E}\{\tilde{x}_u(i, 0)\} = \mathbb{E}\{\tilde{x}_u(0, j)\} = 0.$$

Recalling the statistical properties of the nonlinear function $f(x(i, j), \xi(i, j))$ and the stochastic noise $w(i, j)$ for $i, j \in [0\ N]$, it follows from the dynamics of the one-step prediction error equation (2.6a) that

$$\mathbb{E}\{\tilde{x}_p(1, 1)\} = A_1(1, 0)\mathbb{E}\{\tilde{x}_u(1, 0)\} + A_2(0, 1)\mathbb{E}\{\tilde{x}_u(0, 1)\} = 0. \tag{2.7}$$

Since $\mathbb{E}\{\Xi(i, j) - \bar{\Xi}(i, j)\} = 0$ and $\mathbb{E}\{v(i, j)\} = 0$, it follows from (2.6b) and (2.7) that $\mathbb{E}\{\tilde{x}_u(1, 1)\} = 0$, where the uncorrelatedness between stochastic *variables in (2.1) and stochastic initial values has been utilized. Next, assume that* $\mathbb{E}\{\tilde{x}_u(k,1)\} = 0$ *and* $\mathbb{E}\{\tilde{x}_u(1,l)\} = 0$ *are true for some given constants* k, $l \in [1 \; N-1]$. *The following equalities can be verified from (2.6):*

$$\mathbb{E}\{\tilde{x}_p(k+1,1)\} = A_1(k+1,0)\mathbb{E}\{\tilde{x}_u(k+1,0)\} + A_2(k,1)\mathbb{E}\{\tilde{x}_u(k,1)\} = 0,$$
$$(2.8a)$$

$$\mathbb{E}\{\tilde{x}_p(1,l+1)\} = A_2(0,l+1)\mathbb{E}\{\tilde{x}_u(0,l+1)\} + A_1(1,l)\mathbb{E}\{\tilde{x}_u(1,l)\} = 0,$$
$$(2.8b)$$

$$\mathbb{E}\{\tilde{x}_u(k+1,1)\} = 0, \quad \mathbb{E}\{\tilde{x}_u(1,l+1)\} = 0. \qquad (2.8c)$$

According to the inductive approach, the validity of the following equality holds for all $(i,j) \in \{(k,1)|N \geq k \geq 1\} \cup \{(1,l)|N \geq l > 1\}$:

$$\mathbb{E}\{\tilde{x}_p(i,j)\} = \mathbb{E}\{\tilde{x}_u(i,j)\} = 0. \qquad (2.9)$$

Moreover, suppose that (2.9) is satisfied for all $(i,j) \in \{(i_0,l)|N \geq i_0 \geq k\} \cup \{(k,j_0)|N \geq j_0 > l\}$ *with the given constants* k, $l \in [2 \; N-1]$. *By using the same argument in the derivation of (2.7), one has*

$$\mathbb{E}\{\tilde{x}_p(k+1,l+1)\}$$
$$= A_1(k+1,l)\mathbb{E}\{\tilde{x}_u(k+1,l)\} + A_2(k,l+1)\mathbb{E}\{\tilde{x}_u(k,l+1)\} = 0$$

which, by resorting to the induction method once again, further implies that

$$\mathbb{E}\{\tilde{x}_p(k+1,j)\} = \mathbb{E}\{\tilde{x}_u(k+1,j)\} = 0,$$
$$\mathbb{E}\{\tilde{x}_p(i,l+1)\} = \mathbb{E}\{\tilde{x}_u(i,l+1)\} = 0$$

hold for all j *with* $N \geq j \geq l+1$ *and* i *with* $N \geq i \geq k+1$. *Repeating such a process along with the increments of* i *and* j, *respectively, one can conclude that (2.9) holds for all* $(i,j) \in \mathscr{S}[1 \; N]$. *The proof is now complete.*

According to the statistical characteristics of the random variables in (2.1), Lemma 2.2 provides a rigorous proof to ensure the unbiasedness of filter (2.5). Next, the following two lemmas present the evolutions of the second-order moment of the state and the prediction error variances, respectively.

Lemma 2.3 *For the 2-D system (2.1), the second-order moment of state* $X(i,j)$ *satisfies the following recursion for* i, $j \in [1 \; N]$

$$X(i,j) = A_1(i,j-1)X(i,j-1)A_1^T(i,j-1)$$
$$+ A_2(i-1,j)X(i-1,j)A_2^T(i-1,j)$$
$$+ B_1(i,j-1)Q(i,j-1)B_1^T(i,j-1)$$
$$+ B_2(i-1,j)Q(i-1,j)B_2^T(i-1,j)$$
$$+ \sum_{s=1}^{h} \Pi_s \mathrm{tr}\{(X(i,j-1) + X(i-1,j))\Gamma_s\}$$

$$+ A_1(i, j-1)\mathbb{E}\{x(i, j-1)x^T(i-1, j)\}A_2^T(i-1, j)$$
$$+ A_2(i-1, j)\mathbb{E}\{x(i-1, j)x^T(i, j-1)\}A_1^T(i, j-1). \qquad (2.10)$$

Proof *According to the properties of random variables* $\xi(i, j)$, $w(i, j)$, *and the initial conditions of (2.1), it results immediately from (2.2a) that*

$$\mathbb{E}\{x(i, j)w^T(k, l)\} = 0, \quad \mathbb{E}\{f(x(i, j), \xi(i, j))w^T(k, l)\} = 0$$

for $(k, l) \in \{(k_0, l_0)|k_0 > i \text{ or } l_0 > j\} \cup (i, j)$. *Considering the property of the conditional expectation, it follows from (2.2b) and (2.2c) that*

$$\mathbb{E}\{x(i, j)f^T(x(k, l), \xi(k, l))\}$$
$$= \mathbb{E}\{\mathbb{E}\{x(i, j)f^T(x(k, l), \xi(k, l))|x(i, j)\}\}$$
$$= \mathbb{E}\{x(i, j)\mathbb{E}\{f^T(x(k, l), \xi(k, l))|x(i, j)\}\}$$
$$= 0, \qquad (k, l) \in \{(k_0, l_0)|k_0 > i \text{ or } l_0 > j\} \cup (i, j) \qquad (2.11a)$$
$$\mathbb{E}\{f(x(i, j), \xi(i, j))f^T(x(k, l), \xi(k, l))\}$$
$$= \mathbb{E}\{\mathbb{E}\{f(x(i, j), \xi(i, j))f^T(x(k, l), \xi(k, l))|x(i, j)\}\}$$
$$= \sum_{s=1}^{h} \Pi_s \mathbb{E}\{x^T(i, j)\Gamma_s x(i, j)\}\delta(i, k)\delta(j, l)$$
$$= \sum_{s=1}^{h} \Pi_s \text{tr}\{X(i, j)\Gamma_s\}\delta(i, k)\delta(j, l). \qquad (2.11b)$$

Then, (2.10) can be concluded from (2.1), which completes the proof.

Lemma 2.4 *For the 2-D system (2.1), the variance matrix* $P_p(i, j)$ *of the one-step prediction error has the following recursion for* $i, j \in [1 \ N]$

$$P_p(i, j) = A_1(i, j-1)P_u(i, j-1)A_1^T(i, j-1)$$
$$+ A_2(i-1, j)P_u(i-1, j)A_2^T(i-1, j)$$
$$+ A_1(i, j-1)\mathbb{E}\{\tilde{x}_u(i, j-1)\tilde{x}_u^T(i-1, j)\}A_2^T(i-1, j)$$
$$+ A_2(i-1, j)\mathbb{E}\{\tilde{x}_u(i-1, j)\tilde{x}_u^T(i, j-1)\}A_1^T(i, j-1)$$
$$+ \sum_{s=1}^{h} \Pi_s \text{tr}\{(X(i, j-1) + X(i-1, j))\Gamma_s\}$$
$$+ B_1(i, j-1)Q(i, j-1)B_1^T(i, j-1)$$
$$+ B_2(i-1, j)Q(i-1, j)B_2^T(i-1, j). \qquad (2.12)$$

Proof *From the expression of* $\tilde{x}_u(i, j)$, *it is clear that* $\tilde{x}_u(i, j)$ *is uncorrelated with* $w(k, l)$ *and* $\xi(k, l)$ *if* $(k, l) \in \{(k_0, l_0)|k_0 > i \text{ or } l_0 > j\} \cup (i, j)$. *In addition, since the nonlinear function* $f(x(k, l), \xi(k, l))$ *with* $(k, l) \in \{(k_0, l_0)|$

$k_0 > i$ *or* $l_0 > j\} \cup (i,j)$ *is uncorrelated with the random variables* $v(i_0, j_0)$
and $\{\beta_r(i_0, j_0)\}_{r=1}^m$ *for* $i \geq i_0 \geq 1$ *and* $j \geq j_0 \geq 1$, *we have*

$$
\begin{aligned}
&\mathbb{E}\{\tilde{x}_u(i,j)f^T(x(k,l),\xi(k,l))\} \\
&= \mathbb{E}\{\mathbb{E}\{\tilde{x}_u(i,j)f^T(x(k,l),\xi(k,l))|\tilde{x}_u(i,j)\}\} \\
&= \mathbb{E}\{\tilde{x}_u(i,j)\mathbb{E}\{f^T(x(k,l),\xi(k,l))|\tilde{x}_u(i,j)\}\} \\
&= \mathbb{E}\{\tilde{x}_u(i,j)\mathbb{E}\{f^T(x(k,l),\xi(k,l))|x(i,j)\}\} \\
&= 0.
\end{aligned}
$$

*Along the similar line as in Lemma 2.3, one obtains (2.12) readily and the
detailed proof is omitted here.*

The theorem given below demonstrates that, at each step, the proposed
filter minimizes the filtering error variance.

Theorem 2.1 *For the 2-D system (2.1), the gain parameter* $K(i,j)$ *of filter
(2.5) that minimizes the error variance is determined by*

$$
K(i,j) = P_p(i,j)C^T(i,j)\bar{\Xi}^T(i,j)\bar{R}^{-1}(i,j) \tag{2.13}
$$

where

$$
\begin{aligned}
\bar{R}(i,j) =& \bar{\Xi}(i,j)C(i,j)P_p(i,j)C^T(i,j)\bar{\Xi}(i,j) \\
&+ \Lambda(i,j) \circ (C(i,j)X(i,j)C^T(i,j)) + R(i,j), \\
\Lambda(i,j) =& \mathrm{diag}\{\tilde{\beta}_1(i,j), \tilde{\beta}_2(i,j), \ldots, \tilde{\beta}_m(i,j)\}.
\end{aligned}
$$

Moreover, the minimal filtering error variance is given as

$$
P_u(i,j) = P_p(i,j) - K(i,j)\bar{\Xi}(i,j)C(i,j)P_p(i,j). \tag{2.14}
$$

Proof *Since the stochastic noise* $v(i,j)$ *is uncorrelated with* $\tilde{x}_p(i,j)$ *and*
$x(i,j)$, *one has from (2.6a) that*

$$
\begin{aligned}
P_u(i,j) =& \left[I - K(i,j)\bar{\Xi}(i,j)C(i,j)\right]P_p(i,j)\left[I - K(i,j)\bar{\Xi}(i,j)C(i,j)\right]^T \\
&+ K(i,j)\left[\Lambda(i,j) \circ (C(i,j)X(i,j)C^T(i,j))\right]K^T(i,j) \\
&+ K(i,j)R(i,j)K^T(i,j) \tag{2.15}
\end{aligned}
$$

where Lemma 2.1 has been utilized. To determine the filter parameter $K(i,j)$
minimizing $P_u(i,j)$, *we apply the completing-the-square method to obtain*

$$
\begin{aligned}
P_u(i,j) =& P_p(i,j) - K(i,j)\bar{\Xi}(i,j)C(i,j)P_p(i,j) \\
&- P_p(i,j)C^T(i,j)\bar{\Xi}^T(i,j)K^T(i,j) \\
&+ K(i,j)[\bar{\Xi}(i,j)C(i,j)P_p(i,j)C^T(i,j)\bar{\Xi}(i,j) \\
&+ \Lambda(i,j) \circ (C(i,j)X(i,j)C^T(i,j)) + R(i,j)]K^T(i,j)
\end{aligned}
$$

$$= P_p(i,j) - K_0(i,j)\bar{R}(i,j)K_0^T(i,j)$$
$$+ (K(i,j) - K_0(i,j))\bar{R}(i,j)(K(i,j) - K_0(i,j))^T, \qquad (2.16)$$

where $K_0(i,j) = P_p(i,j)C^T(i,j)\bar{\Xi}^T(i,j)\bar{R}^{-1}(i,j)$. *The filtering error covariance* $P_u(i,j)$ *is minimized if and only if* $K(i,j) = K_0(i,j)$, *which infers the validity of (2.13).*

Substituting (2.13) into (2.16), the minimum error variance can be calculated as

$$P_u(i,j) = P_p(i,j) - K_0(i,j)\bar{R}(i,j)K_0^T(i,j)$$
$$= P_p(i,j) - K(i,j)\bar{\Xi}(i,j)C(i,j)P_p(i,j)$$

which completes the proof.

The optimal filter derived above minimizes the filtering error variance. In the next section, we will analyze the performance of the obtained filter.

2.3 Performance Analysis

In this section, the performance of the designed filter is analyzed. More specifically, an upper bound of the filtering error variance is first given in the sense of spectral norm, and then the monotonicity with respect to the degradation probability of the measurement is discussed. Finally, a computational algorithm is presented for implementing the designed filters.

2.3.1 Boundedness Analysis

To proceed, the following assumption on the system parameters is needed for the boundedness analysis.

Assumption 2.1 *There are positive scalars* $\bar{a}_l$, $\bar{b}_l$ $(l = 1, 2)$, $\bar{\gamma}_s$, $\bar{\pi}_s$ $(s \in [1\ h])$, $\bar{m}$, *and* $\bar{q}$ *such that the following inequalities are satisfied for* $i, j \in [0\ N]$:

$$\|A_l(i,j)\| \le \bar{a}_l, \quad \|B_l(i,j)\| \le \bar{b}_l, \quad \|\Gamma_s + \Gamma_s^T\| \le \bar{\gamma}_s,$$
$$\|\Pi_s\| \le \bar{\pi}_s, \quad \mathrm{tr}\{X(i,j)\} \le \bar{m}, \quad \|Q(i,j)\| \le \bar{q}.$$

Theorem 2.2 *Consider the 2-D stochastic system (2.1) with filter (2.5) whose gain parameter* $K(i,j)$ *is presented in Theorem 2.1. Under Assumption 1, the minimal filtering error variance (2.14) is bounded by*

$$\|P_u(i,j)\| \le \sum_{k=1}^{i}(1+\mu)\bar{a}_1^2 r(i-k, j-1)\|P_u(k,0)\|$$

$$+ \sum_{l=1}^{j}(1 + \mu^{-1})\bar{a}_2^2 r(i - 1, j - l)\|P_u(0, l)\|$$

$$+ \sum_{k=0}^{i-1}\sum_{l=0}^{j-1} r(i - k - 1, j - l - 1)b_0, \qquad (2.17)$$

where $\mu > 0$ is an arbitrary scalar, $b_0 = \sum_{s=1}^{h} \bar{m}\bar{\gamma}_s\bar{\pi}_s + (\bar{b}_1^2 + \bar{b}_2^2)\bar{q}$, and $r(\cdot, \cdot)$ with $r(0, 0) = 1$ is recursively determined by

$$r(0, j) = (1 + \mu)\bar{a}_1^2 r(0, j - 1), \qquad (2.18a)$$

$$r(i, 0) = (1 + \mu^{-1})\bar{a}_2^2 r(i - 1, 0), \qquad (2.18b)$$

$$r(i, j) = (1 + \mu)\bar{a}_1^2 r(i, j - 1) + (1 + \mu^{-1})\bar{a}_2^2 r(i - 1, j), \quad i, j \in [1 \ N]. \quad (2.18c)$$

Proof *For any given constant $\mu > 0$, it follows from (2.12) that*

$$\begin{aligned} P_p(i, j) \leq &(1 + \mu)A_1(i, j - 1)P_u(i, j - 1)A_1^T(i, j - 1) \\ &+ (1 + \mu^{-1})A_2(i - 1, j)P_u(i - 1, j)A_2^T(i - 1, j) \\ &+ \sum_{s=1}^{h}\Pi_s\mathrm{tr}\{(X(i, j - 1) + X(i - 1, j))\Gamma_s\} \\ &+ B_1(i, j - 1)Q(i, j - 1)B_1^T(i, j - 1) \\ &+ B_2(i - 1, j)Q(i - 1, j)B_2^T(i - 1, j). \end{aligned} \qquad (2.19)$$

It is evident from Assumption 2.1 that

$$\begin{aligned} &\mathrm{tr}\{(X(i, j - 1) + X(i - 1, j))\Gamma_s\} \\ &= \mathbb{E}\{x^T(i, j - 1)\Gamma_s x(i, j - 1) + x^T(i - 1, j)\Gamma_s x(i - 1, j)\} \\ &= \frac{1}{2}\mathbb{E}\left\{x^T(i, j - 1)(\Gamma_s^T + \Gamma_s)x(i, j - 1) + x^T(i - 1, j)(\Gamma_s^T + \Gamma_s)x(i - 1, j)\right\} \\ &\leq \frac{1}{2}\bar{\gamma}_s\mathbb{E}\{x^T(i, j - 1)x(i, j - 1) + x^T(i - 1, j)x(i - 1, j)\} \\ &\leq \bar{m}\bar{\gamma}_s. \end{aligned} \qquad (2.20)$$

Substituting (2.20) into (2.19), one obtains

$$\|P_p(i, j)\| \leq (1 + \mu)\bar{a}_1^2\|P_u(i, j - 1)\| + (1 + \mu^{-1})\bar{a}_2^2\|P_u(i - 1, j)\| + b_0.$$

Furthermore, noticing that $\|P_u(i, j)\| \leq \|P_p(i, j)\|$ is true according to (2.13) and (2.14), we have

$$\|P_u(i, j)\| \leq (1 + \mu)\bar{a}_1^2\|P_u(i, j - 1)\| + (1 + \mu^{-1})\bar{a}_2^2\|P_u(i - 1, j)\| + b_0 \quad (2.21)$$

which confirms that (2.17) is true for $(i, j) = (1, 1)$.

Next, to prove this theorem by induction, we assume that (2.17) holds for

$(i_0, 1)$ and $(1, j_0)$ where $i_0, j_0 \in [2\ N-1]$. It is deduced from (2.21) and the inductive hypothesis that

$$\|P_u(i_0+1,1)\| \leq (1+\mu)\bar{a}_1^2\|P_u(i_0+1,0)\| + (1+\mu^{-1})\bar{a}_2^2\|P_u(i_0,1)\| + b_0$$

$$\leq (1+\mu)\bar{a}_1^2\|P_u(i_0+1,0)\|$$

$$+ (1+\mu^{-1})\bar{a}_2^2\Big\{\sum_{k=1}^{i_0}(1+\mu)\bar{a}_1^2 r(i_0-k,0)\|P_u(k,0)\|$$

$$+ (1+\mu^{-1})\bar{a}_2^2 r(i_0-1,0)\|P_u(0,1)\|$$

$$+ \sum_{k=0}^{i_0-1} r(i_0-k-1,0)b_0\Big\} + b_0$$

$$= (1+\mu)\bar{a}_1^2\|P_u(i_0+1,0)\|$$

$$+ \sum_{k=1}^{i_0}(1+\mu)\bar{a}_1^2 r(i_0-k+1,0)\|P_u(k,0)\|$$

$$+ (1+\mu^{-1})\bar{a}_2^2 r(i_0,0)\|P_u(0,1)\| + \Big\{\sum_{k=0}^{i_0-1} r(i_0-k,0)+1\Big\}b_0$$

$$= \sum_{k=1}^{i_0+1}(1+\mu)\bar{a}_1^2 r(i_0-k+1,0)\|P_u(k,0)\|$$

$$+ (1+\mu^{-1})\bar{a}_2^2 r(i_0,0)\|P_u(0,1)\| + \sum_{k=0}^{i_0} r(i_0-k,0)b_0 \quad (2.22)$$

where the iterative definition (2.18) for function $r(\cdot,\cdot)$ has been utilized. Similarly, one has

$$\|P_u(1,j_0+1)\| \leq (1+\mu)\bar{a}_1^2\|P_u(1,j_0)\| + (1+\mu^{-1})\bar{a}_2^2\|P_u(0,j_0+1)\| + b_0$$

$$\leq (1+\mu)\bar{a}_1^2 r(0,j_0)\|P_u(1,0)\|$$

$$+ \sum_{l=1}^{j_0}(1+\mu^{-1})\bar{a}_2^2 r(0,j_0-l+1)\|P_u(0,l)\|$$

$$+ (1+\mu^{-1})\bar{a}_2^2\|P_u(0,j_0+1)\| + \Big\{\sum_{l=0}^{j_0-1} r(0,j_0-l)+1\Big\}b_0$$

$$= (1+\mu)\bar{a}_1^2 r(0,j_0)\|P_u(1,0)\| + \sum_{l=0}^{j_0} r(0,j_0-l)b_0$$

$$+ \sum_{l=1}^{j_0+1}(1+\mu^{-1})\bar{a}_2^2 r(0,j_0-l+1)\|P_u(0,l)\|. \quad (2.23)$$

Equalities (2.22) and (2.23) infer that (2.17) holds for $(i_0+1,1)$ and $(1,j_0+$

1). *Therefore, the induction method assures that (2.17) is valid for* $(i, j) \in \{(i_0, 1)|N \geq i_0 \geq 1\} \cup \{(1, j_0)|N \geq j_0 > 1\}$.

Finally, assume that (2.17) is valid for $(i, j) \in \{(i_1, j_0)|N \geq i_1 \geq i_0\} \cup \{(i_0, j_1)|j_1 > j_0\}$ *with given constants* $i_0, j_0 \in [2\ N-1]$. *Then, it follows from (2.21), the inductive hypothesis, and the definition of* $r(\cdot, \cdot)$ *in (2.18) that*

$$
\begin{aligned}
&\|P_u(i_0 + 1, j_0 + 1)\| \\
&\leq (1+\mu)\bar{a}_1^2 \|P_u(i_0+1, j_0)\| + (1+\mu^{-1})\bar{a}_2^2 \|P_u(i_0, j_0+1)\| + b_0 \\
&\leq (1+\mu)\bar{a}_1^2 \Big\{ \sum_{k=1}^{i_0+1} (1+\mu)\bar{a}_1^2 r(i_0 - k + 1, j_0 - 1)\|P_u(k, 0)\| \\
&\quad + \sum_{l=1}^{j_0} (1+\mu^{-1})\bar{a}_2^2 r(i_0, j_0 - l)\|P_u(0, l)\| \\
&\quad + \sum_{k=0}^{i_0} \sum_{l=0}^{j_0-1} r(i_0 - k, j_0 - l - 1)b_0 \Big\} \\
&\quad + (1+\mu^{-1})\bar{a}_2^2 \Big\{ \sum_{k=1}^{i_0} (1+\mu)\bar{a}_1^2 r(i_0 - k, j_0)\|P_u(k, 0)\| \\
&\quad + \sum_{l=1}^{j_0+1} (1+\mu^{-1})\bar{a}_2^2 r(i_0 - 1, j_0 - l + 1)\|P_u(0, l)\| \\
&\quad + \sum_{k=0}^{i_0-1} \sum_{l=0}^{j_0} r(i_0 - k - 1, j_0 - l)b_0 \Big\} + b_0 \\
&= \sum_{k=1}^{i_0} (1+\mu)\bar{a}_1^2 r(i_0 - k + 1, j_0)\|P_u(k, 0)\| \\
&\quad + (1+\mu)\bar{a}_1^2 r(0, j_0)\|P_u(i_0 + 1, 0)\| \\
&\quad + \sum_{l=1}^{j_0} (1+\mu^{-1})\bar{a}_2^2 r(i_0, j_0 - l + 1)\|P_u(0, l)\| \\
&\quad + (1+\mu^{-1})\bar{a}_2^2 r(i_0, 0)\|P_u(0, j_0 + 1)\| + \Big\{ \sum_{k=0}^{i_0-1} \sum_{l=0}^{j_0-1} r(i_0 - k, j_0 - l) \\
&\quad + \sum_{l=0}^{j_0-1} r(0, j_0 - l) + \sum_{k=0}^{i_0-1} r(i_0 - k, 0) + 1 \Big\} b_0 \\
&= \sum_{k=1}^{i_0+1} (1+\mu)\bar{a}_1^2 r(i_0 - k + 1, j_0)\|P_u(k, 0)\| + \sum_{k=0}^{i_0} \sum_{l=0}^{j_0} r(i_0 - k, j_0 - l)b_0 \\
&\quad + \sum_{l=1}^{j_0+1} (1+\mu^{-1})\bar{a}_2^2 r(i_0, j_0 - l + 1)\|P_u(0, l)\|
\end{aligned}
$$

which means that (2.17) is true for $(i,j) = (i_0 + 1, j_0 + 1)$. *Similar to the derivation of (2.22) and (2.23), some tedious but straightforward manipulations yield (2.17) for* $(i,j) \in \{(i_1, j_0+1)|i_1 \geq i_0+1\} \cup \{(i_0+1, j_1)|j_1 > j_0+1\}$, *and the proof is now complete via the induction approach.*

Remark 2.4 *For the 2-D stochastic system (2.1) with filter (2.5), under Assumption 2.1, Theorem 2.2 provides the boundedness analysis (in the sense of spectral norm) for the minimal filtering error variance* $P_u(i,j)$. *From the practical point of view, due to the energy constraints over a finite horizon, it is quite reasonable to assume that the spectral norms of system matrices, the trace of the second-order moment of the state and the process noise covariance are all bounded. Then, by means of mathematical induction, a recursive method is utilized to quantify the bound of* $P_u(i,j)$. *It can be seen from (2.17) that the developed bound consists of the spectral norms of system parameters, the noise covariance as well as the estimation error-variances along boundary, thereby reflecting all necessary information on the error dynamics.*

2.3.2 Monotonicity Analysis

Note that the filtering error covariance $\mathrm{tr}\{P_u(i,j)\}$ serves as the most important index for the filtering performance. Let us now discuss the relationship between the measurement degradation and the filtering performance to evaluate the optimal filter established in Theorem 2.1. For the convenience of illustration, the random variables $\beta_r(i,j)$ $(r \in [1 \ m])$ are supposed to be unified as $\beta(i,j)$ whose expectations and variances are denoted as $\bar{\beta}(i,j)$ and $\tilde{\beta}(i,j)$, respectively. The following lemma is introduced for later development.

Lemma 2.5 *[154] For derivable matrix functions F and H with respect to variable* $x \in \mathbb{R}$, *where H is also invertible, the following equalities hold*

$$\frac{d\mathrm{tr}\{F\}}{dx} = \mathrm{tr}\left\{\frac{dF}{dx}\right\}, \quad \frac{dH^{-1}}{dx} = -H^{-1}\frac{dH}{dx}H^{-1}.$$

The following two theorems show that the degradation probability $\bar{\beta}(i,j)$ indeed influences the performance index $\mathrm{tr}\{P_u(i,j)\}$.

Theorem 2.3 *Let the variance* $\tilde{\beta}(i,j)$ *be fixed whereas* $\bar{\beta}(i,j)$ *be variable. For the optimal filter (2.5), the performance index* $\mathrm{tr}\{P_u(i,j)\}$ *is non-increasing with increased* $\bar{\beta}(i,j)$.

Proof *According to (2.13) and (2.14), the filtering error variance* $P_u(i,j)$ *can be rewritten as*

$$P_u(i,j) = P_p(i,j) - P_p(i,j)C^T(i,j)\bar{\Xi}^T(i,j)\bar{R}^{-1}(i,j)\bar{\Xi}(i,j)C(i,j)P_p(i,j)$$

and then it follows from Lemma 2.5 that

$$\frac{d\mathrm{tr}\{P_u(i,j)\}}{d\bar{\beta}(i,j)} = \mathrm{tr}\left\{\frac{dP_u(i,j)}{d\bar{\beta}(i,j)}\right\}$$

$$\begin{aligned}
&= \operatorname{tr}\Big\{ -2P_p(i,j)C^T(i,j)\bar{\Xi}^T(i,j)\bar{R}^{-1}(i,j)C(i,j)P_p(i,j) \\
&\quad + 2\bar{\beta}(i,j)P_p(i,j)C^T(i,j)\bar{\Xi}^T(i,j)\bar{R}^{-1}(i,j)C(i,j)P_p(i,j) \\
&\quad \times C^T(i,j)\bar{R}^{-1}(i,j)\bar{\Xi}(i,j)C(i,j)P_p(i,j) \Big\} \\
&= \frac{2}{\bar{\beta}(i,j)}\operatorname{tr}\Big\{ -K(i,j)\bar{R}(i,j)K(i,j) \\
&\quad + K(i,j)\bar{\Xi}(i,j)C(i,j)P_p(i,j)C^T(i,j)\bar{\Xi}^T(i,j)K^T(i,j) \Big\} \\
&= -\frac{2}{\bar{\beta}(i,j)}\operatorname{tr}\Big\{ K(i,j)[\Lambda(i,j)\circ(C(i,j)X(i,j)C^T(i,j)) + R(i,j)]K^T(i,j) \Big\}.
\end{aligned}$$

Since $\bar{\beta}(i,j) > 0$ and the trace of a positive semi-definite matrix is always non-negative, one has

$$\frac{d\operatorname{tr}\{P_u(i,j)\}}{d\bar{\beta}(i,j)} \le 0$$

which ends the proof.

Theorem 2.4 *Let the random variable $\beta(i,j)$ be a Bernoulli distributed white sequence taking values on 0 or 1 with*

$$\operatorname{Prob}\{\beta(i,j)=1\} = \bar{\beta}(i,j), \qquad \operatorname{Prob}\{\beta(i,j)=0\} = 1 - \bar{\beta}(i,j). \qquad (2.24)$$

In this case, for the optimal filter (2.5), the performance index $\operatorname{tr}\{P_u(i,j)\}$ is non-increasing with increased $\bar{\beta}(i,j)$.

Proof *It follows from (2.24) that $\bar{\beta}(i,j) = \mathbb{E}\{\beta(i,j)\}$ and $\tilde{\beta}(i,j) = \bar{\beta}(i,j) - \bar{\beta}^2(i,j)$. In this case, the variance $\tilde{\beta}(i,j)$ is a function of $\bar{\beta}(i,j)$ and it is obvious that matrix $\bar{R}(i,j)$ can be expressed as follows:*

$$\begin{aligned}
\bar{R}(i,j) &= R(i,j) + \bar{\beta}^2(i,j)C(i,j)P_p(i,j)C^T(i,j) \\
&\quad + \bar{\beta}(i,j)\left(1 - \bar{\beta}(i,j)\right) I \circ \left(C(i,j)X(i,j)C^T(i,j)\right).
\end{aligned}$$

Recalling the expression of $P_u(i,j)$, the following equality is valid by exploiting Lemma 2.5:

$$\begin{aligned}
\frac{d\operatorname{tr}\{P_u(i,j)\}}{d\bar{\beta}(i,j)} &= \operatorname{tr}\left\{\frac{dP_u(i,j)}{d\bar{\beta}(i,j)}\right\} \\
&= \operatorname{tr}\Big\{ -2\bar{\beta}(i,j)P_p(i,j)C^T(i,j)\bar{R}^{-1}(i,j)C(i,j)P_p(i,j) \\
&\quad + \bar{\beta}^2(i,j)P_p(i,j)C^T(i,j)\bar{R}^{-1}(i,j)\Big[2\bar{\beta}(i,j)C(i,j)P_p(i,j)C^T(i,j) \\
&\quad + \left(1 - 2\bar{\beta}(i,j)\right)I \circ \left(C(i,j)X(i,j)C^T(i,j)\right)\Big]\bar{R}^{-1}(i,j)C(i,j)P_p(i,j) \Big\} \\
&= -\bar{\beta}^{-1}(i,j)\operatorname{tr}\Big\{ K(i,j)\Big[2\bar{R}(i,j) - 2\bar{\beta}^2(i,j)C(i,j)P_p(i,j)C^T(i,j)
\end{aligned}$$

$$- \bar{\beta}(i,j)\big(1 - 2\bar{\beta}(i,j)\big)I \circ \big(C(i,j)X(i,j)C^T(i,j)\big)\Big]K^T(i,j)\Big\}$$

$$= -\bar{\beta}^{-1}(i,j)\mathrm{tr}\Big\{K(i,j)\Big[\bar{\beta}(i,j)I \circ \big(C(i,j)X(i,j)C^T(i,j)\big)$$

$$+ 2R(i,j)\Big]K^T(i,j)\Big\}. \tag{2.25}$$

Based on the property of the positive semi-definite matrix, it is clear that

$$\frac{d\mathrm{tr}\{P_u(i,j)\}}{d\bar{\beta}(i,j)} \leq 0$$

which completes the proof.

Remark 2.5 *When $\bar{\Xi}(i,j) = \bar{\beta}(i,j)I$, Theorems 2.3 and 2.4 demonstrate the monotonicity of $\mathrm{tr}\{P_u(i,j)\}$ with respect to $\bar{\beta}(i,j)$ in two different cases. Since the probability $\bar{\beta}(i,j)$ is introduced to reflect the phenomenon of degraded measurements, a greater value of $\bar{\beta}(i,j)$ means that the filter performs under a better working condition and, with more available measurements, a better filtering performance (i.e., a lower value of $\mathrm{tr}\{P_u(i,j)\}$) can be expected, which conforms with the practical situation.*

2.3.3 Filtering Algorithm

After the design of the optimal filter and the analysis of its performance, we are ready to present some algorithms for designing the optimal filter (2.5).

As can be seen in (2.10) and (2.12), the evolutions of $X(i,j)$ and $P_p(i,j)$ involve, respectively, the cross-terms $\mathbb{E}\{x(i,j-1)x^T(i-1,j)\}$ and $\mathbb{E}\{\tilde{x}_u(i,j-1)\tilde{x}_u^T(i-1,j)\}$, whose recursions should be further derived. By utilizing the statistical properties of random variables considered in system (2.1), for $i,j \in [2\ N]$, one obtains the following two recursions:

$$\mathbb{E}\left\{x(i,j-1)x^T(i-1,j)\right\}$$

$$= A_1(i,j-2)\mathbb{E}\left\{x(i,j-2)x^T(i-1,j-1)\right\}A_1^T(i-1,j-1)$$

$$+ A_1(i,j-2)\mathbb{E}\left\{x(i,j-2)x^T(i-2,j)\right\}A_2^T(i-2,j)$$

$$+ A_2(i-1,j-1)\mathbb{E}\left\{x(i-1,j-1)x^T(i-2,j)\right\}A_2^T(i-2,j)$$

$$+ A_2(i-1,j-1)X(i-1,j-1)A_1^T(i-1,j-1)$$

$$+ \sum_{s=1}^{h}\Pi_s\mathrm{tr}\{X(i-1,j-1)\Gamma_s\}$$

$$+ B_2(i-1,j-1)Q(i-1,j-1)B_1^T(i-1,j-1), \tag{2.26}$$

$$\mathbb{E}\left\{\tilde{x}_u(i,j-1)\tilde{x}_u^T(i-1,j)\right\}$$

$$= \big[I - K(i,j-1)\bar{\Xi}(i,j-1)C(i,j-1)\big]\mathbb{E}\left\{\tilde{x}_p(i,j-1)\tilde{x}_p^T(i-1,j)\right\}$$

$$\times \big[I - K(i-1,j)\bar{\Xi}(i-1,j)C(i-1,j)\big]^T$$

$$= \left[I - K(i,j-1)\bar{\Xi}(i,j-1)C(i,j-1) \right]$$

$$\times \left\{ A_1(i,j-2)\mathbb{E}\left\{ \tilde{x}_u(i,j-2)\tilde{x}_u^T(i-1,j-1) \right\} A_1^T(i-1,j-1) \right.$$

$$+ A_1(i,j-2)\mathbb{E}\left\{ \tilde{x}_u(i,j-2)\tilde{x}_u^T(i-2,j) \right\} A_2^T(i-2,j)$$

$$+ A_2(i-1,j-1)P_u(i-1,j-1)A_1^T(i-1,j-1)$$

$$+ A_2(i-1,j-1)\mathbb{E}\left\{ \tilde{x}_u(i-1,j-1)\tilde{x}_u^T(i-2,j) \right\} A_2^T(i-2,j)$$

$$+ \sum_{s=1}^{h} \Pi_s \mathrm{tr}\{ X(i-1,j-1)\Gamma_s \}$$

$$+ \left. B_2(i-1,j-1)Q(i-1,j-1)B_1^T(i-1,j-1) \right\}$$

$$\times \left[I - K(i-1,j)\bar{\Xi}(i-1,j)C(i-1,j) \right]^T . \tag{2.27}$$

Similarly, for $i,j \in [k\ N]$ and $k \in [2\ N-1]$, one has the following recursions:

$$\mathbb{E}\left\{ x(i,j-k)x^T(i-k,j) \right\}$$

$$= A_1(i,j-k-1)\mathbb{E}\left\{ x(i,j-k-1)x^T(i-k,j-1) \right\} A_1^T(i-k,j-1)$$

$$+ A_1(i,j-k-1)\mathbb{E}\left\{ x(i,j-k-1)x^T(i-k-1,j) \right\} A_2^T(i-k-1,j)$$

$$+ A_2(i-1,j-k)\mathbb{E}\left\{ x(i-1,j-k)x^T(i-k,j-1) \right\} A_1^T(i-k,j-1)$$

$$+ A_2(i-1,j-k)\mathbb{E}\left\{ x(i-1,j-k)x^T(i-k-1,j) \right\} A_2^T(i-k-1,j), \tag{2.28}$$

$$\mathbb{E}\left\{ \tilde{x}_u(i,j-k)\tilde{x}_u^T(i-k,j) \right\}$$

$$= \left[I - K(i,j-k)\bar{\Xi}(i,j-k)C(i,j-k) \right] \mathbb{E}\left\{ \tilde{x}_p(i,j-k)\tilde{x}_p^T(i-k,j) \right\}$$

$$\times \left[I - K(i-k,j)\bar{\Xi}(i-k,j)C(i-k,j) \right]^T$$

$$= \left[I - K(i,j-k)\bar{\Xi}(i,j-k)C(i,j-k) \right]$$

$$\times \left\{ A_1(i,j-k-1)\mathbb{E}\left\{ \tilde{x}_u(i,j-k-1)\tilde{x}_u^T(i-k,j-1) \right\} A_1^T(i-k,j-1) \right.$$

$$+ A_1(i,j-k-1)\mathbb{E}\left\{ \tilde{x}_u(i,j-k-1)\tilde{x}_u^T(i-k-1,j) \right\} A_2^T(i-k-1,j)$$

$$+ A_2(i-1,j-k)\mathbb{E}\left\{ \tilde{x}_u(i-1,j-k)\tilde{x}_u^T(i-k,j-1) \right\} A_1^T(i-k,j-1)$$

$$+ \left. A_2(i-1,j-k)\mathbb{E}\left\{ \tilde{x}_u(i-1,j-k)\tilde{x}_u^T(i-k-1,j) \right\} A_2^T(i-k-1,j) \right\}$$

$$\times \left[I - K(i-k,j)\bar{\Xi}(i-k,j)C(i-k,j) \right]^T . \tag{2.29}$$

According to the recursions (2.5), (2.10), (2.12)–(2.14), and (2.26)–(2.29) that we have acquired so far, we are able to outline an overall algorithm for the 2-D filter design that can be implemented iteratively with given initial conditions. The Kalman-type filtering (KTF) algorithm is summarized as shown in Algorithm 1.

Algorithm 1 KTF

Initialization: Give $u_1(i)$, $u_2(j)$, $P(i,0)$, and $P(0,j)$ for all $i,j \in [0 \ N]$.

1: Set $k = 1$ and $l = 1$

2: **while** $k \in [1 \ N]$ and $l \in [1 \ N]$ **do**

3:　　Compute first $\hat{x}_p(k,l)$, $X(k,l)$, and $P_p(k,l)$ from (2.5a), (2.10), and (2.12), respectively; obtain secondly the gain matrix $K(k,l)$, the parameters $\hat{x}_u(k,l)$, and $P_u(k,l)$ from (2.13), (2.5b), and (2.14), respectively.

4:　　**if** $k \in [1 \ N]$ and $l \in [1 \ N-1]$ **then**

5:　　　　Solve equations (2.26)–(2.29) to obtain the cross-terms $\mathbb{E}\{x(k,l)x^T(i_0, k + l - i_0)\}$ and $\mathbb{E}\{\tilde{x}_u(k,l)\tilde{x}_u^T(i_0, k + l - i_0)\}$ ($i_0 \in [k+l - \min\{k+l, N\} \quad k-1]$).

6:　　**end if**

7:　　**if** $k \in [1 \ N]$ and $l = N$ **then**

8:　　　　Set $k = k + 1$ and $l = 1$.

9:　　**else**

10:　　　　Set $l = l + 1$.

11:　　**end if**

12: **end while**

Remark 2.6 *For the minimum-variance filter design problem of the addressed 2-D system (2.1), a recursive processing has been established in (2.5), (2.10), (2.12)–(2.14), and (2.26)–(2.29), through which the gain parameter $K(\cdot,\cdot)$, the state estimation $\hat{x}_u(\cdot,\cdot)$, and the minimal filtering error variance $P_u(\cdot,\cdot)$ can be calculated at each shifting step. Compared with the traditional 1-D Kalman filtering, the 2-D KTF algorithm is much more complex due to the two-directional signal evolution. To implement the algorithm proposed in this chapter, firstly, fix an index i and solve the recursions iteratively with j varying from 1 to N to obtain certain parameters; secondly, increase the index i to $i + 1$; and then calculate the desired parameters again along the j axis (i.e., j varies from 1 to N). Repeating such a procedure over the finite horizon $[1 \ N] \times [1 \ N]$ iteratively, the above algorithm provides an executable/effective way to accomplish the filter design scheme.*

Remark 2.7 *As a crucial matter in practical applications, the computational burden will be further discussed in the following for the proposed filtering algorithm. With the aid of arithmetic operations in acquiring the estimate vectors $\hat{x}_p(i,j)$ and $\hat{x}_u(i,j)$, matrices $X(i,j)$, $P_p(i,j)$, and $P_u(i,j)$ as well as the filter gain $K(i,j)$, the whole computational complexity of the KTF algorithm over the finite horizon $[0 \ N] \times [0 \ N]$ is computed as $O((n^3 h + n^2(p + m) + n(p^2 + m^2) + m^3)N^2)$.*

Remark 2.8 *This chapter presents one of the first few results on 2-D minimum-variance filtering problem. Rather than considering the steady-state dynamics, this chapter concentrates on investigating the transient behaviors associated with the addressed system over a finite-horizon, which is certainly* a compelling yet challenging issue on account of the two-directional dynamic evolution, the shift-varying feature, the stochastic nonlinearity and the measurement degradation. A recursive filtering strategy is first designed, and evaluation of the filtering performance is then discussed with rigorous derivations. For the obtained results, besides their theoretical significance, promising potentials are expected as they can provide some reference for monitoring certain industrial processes.

2.4   Numerical Example

A numerical example is provided in this section to illustrate the effectiveness and applicability of the developed optimal filtering scheme. Consider a long transmission line in circuit systems. The relationship between its voltage and current is described as follows [83, 126, 132]:

$$\frac{\partial U(z,t)}{\partial z} = L_l \frac{\partial I(z,t)}{\partial t}, \qquad \frac{\partial I(z,t)}{\partial z} = C_c \frac{\partial U(z,t)}{\partial t}, \qquad (2.30)$$

where $U(z,t)$ is the voltage and $I(z,t)$ is the current at space $z \in [0,D]$ with D denoting the length of the line and time $t \in [0,T]$ representing a certain duration range, L_l and C_c are the line inductance and capacitance, respectively.

By defining

$$\begin{bmatrix} \bar{U}(z,t) \\ \bar{I}(z,t) \end{bmatrix} = 0.1 \begin{bmatrix} 1 & \sqrt{\frac{L_l}{C_c}} \\ \sqrt{\frac{C_c}{L_l}} & -1 \end{bmatrix} \begin{bmatrix} U(z,t) \\ I(z,t) \end{bmatrix}$$

we have the following equation from (2.30):

$$\frac{\partial}{\partial t} \begin{bmatrix} \bar{U}(z,t) \\ \bar{I}(z,t) \end{bmatrix} = \begin{bmatrix} \frac{1}{\sqrt{L_l C_c}} & 0 \\ 0 & -\frac{1}{\sqrt{L_l C_c}} \end{bmatrix} \frac{\partial}{\partial z} \begin{bmatrix} \bar{U}(z,t) \\ \bar{I}(z,t) \end{bmatrix}. \qquad (2.31)$$

To discretize (2.31), we assume that $\bar{U}(z,t) \triangleq U_d(i\Delta z, j\Delta t)$ and $\bar{I}(z,t) \triangleq I_d(i\Delta z, j\Delta t)$ for $z \in [i\Delta z, (i+1)\Delta z)$ and $t \in [j\Delta t, (j+1)\Delta t)$. Note that such an assumption is reasonable when choosing appropriate step sizes Δz and Δt since smaller step sizes will result in a higher approximation to the PDE (2.31). Moreover, the following approximate equalities are employed [83]:

$$\frac{\partial \bar{U}(z,t)}{\partial z} \doteq \frac{U_d(i\Delta z, j\Delta t) - U_d((i-1)\Delta z, j\Delta t)}{\Delta z}, \qquad (2.32)$$

$$\frac{\partial \bar{U}(z,t)}{\partial t} \doteq \frac{U_d(i\Delta z, (j+1)\Delta t) - U_d(i\Delta z, j\Delta t)}{\Delta t}, \qquad (2.33)$$

$$\frac{\partial \bar{I}(z,t)}{\partial z} \doteq \frac{I_d(i\Delta z, j\Delta t) - I_d((i-1)\Delta z, j\Delta t)}{\Delta z}, \tag{2.34}$$

$$\frac{\partial \bar{I}(z,t)}{\partial t} \doteq \frac{I_d(i\Delta z, (j+1)\Delta t) - I_d(i\Delta z, j\Delta t)}{\Delta t}. \tag{2.35}$$

For presentation simplification, we denote

$$U_d(i,j) \triangleq U_d(i\Delta z, j\Delta t), \quad I_d(i,j) \triangleq I_d(i\Delta z, j\Delta t),$$
$$x_d^h(i,j) \triangleq [U_d(i-1,j) \ I_d(i-1,j)]^T,$$
$$x_d^v(i,j) \triangleq [U_d(i,j) \ I_d(i,j)]^T,$$
$$x(i,j) \triangleq [x_d^h(i-1,j) \ x_d^v(i-1,j)]^T.$$

Substituting (2.32)–(2.34) into (2.31) yields

$$U_d(i,j+1) = (1 + \frac{\Delta t}{\Delta z}\frac{1}{\sqrt{L_l C_c}})U_d(i,j) - \frac{\Delta t}{\Delta z}\frac{1}{\sqrt{L_l C_c}}U_d(i-1,j),$$

$$I_d(i,j+1) = (1 - \frac{\Delta t}{\Delta z}\frac{1}{\sqrt{L_l C_c}})I_d(i,j) + \frac{\Delta t}{\Delta z}\frac{1}{\sqrt{L_l C_c}}I_d(i-1,j)$$

and therefore

$$x(i+1,j+1) = A_1 x(i+1,j) + A_2 x(i,j+1),$$

where

$$A_1 = \begin{bmatrix} 0 & 0 & 0 & 0 \\ 0 & 0 & 0 & 0 \\ -\wp & 0 & 1+\wp & 0 \\ 0 & \wp & 0 & 1-\wp \end{bmatrix}, \quad A_2 = \begin{bmatrix} 0 & 0 & 1 & 0 \\ 0 & 0 & 0 & 1 \\ 0 & 0 & 0 & 0 \\ 0 & 0 & 0 & 0 \end{bmatrix}$$

with $\wp = (\Delta t/\Delta z)(1/\sqrt{L_l C_c})$.

We now take the shift-varying parameter perturbations, the stochastic disturbances and the degraded measurements into account. Setting the step sizes as $\Delta z = \Delta t = 0.02$, the following 2-D system is obtained on a finite horizon $\mathscr{S}[1 \ N]$ with $N = 50$:

$$x(i,j) = A_1(i,j-1)x(i,j-1) + A_2(i-1,j)x(i-1,j)$$
$$+ f(x(i,j-1), \xi(i,j-1)) + f(x(i-1,j), \xi(i-1,j))$$
$$+ B_1(i,j-1)w(i,j-1) + B_2(i-1,j)w(i-1,j), \tag{2.36a}$$
$$y(i,j) = \beta(i,j)C(i,j)x(i,j) + v(i,j), \tag{2.36b}$$

where

$$A_1(i,j) = \begin{bmatrix} 0 & 0 & 0 & 0 \\ 0 & 0 & 0 & 0 \\ -a(i,j) & 0 & 1+a(i,j) & 0 \\ 0 & a(i,j) & 0 & 1-a(i,j) \end{bmatrix},$$

$$A_2(i,j) = A_2, \quad a(i,j) = \frac{1}{\sqrt{L_l(i,j)C_c(i,j)}},$$

$$C_c(i,j) = (30 - \sin(i)\cos(j))F/mt,$$

$$L_l(i,j) = (30 + e^{-i})H/mt,$$

$$B_1(i,j) = \begin{bmatrix} 0.03 & 0.01 & 0.02 + 0.01e^{-5i} & 0 \end{bmatrix},$$

$$B_2(i,j) = \begin{bmatrix} 0.05 - 0.02\cos(j) & 0 & -0.03 & 0 \end{bmatrix}^T,$$

$$C(i,j) = \begin{bmatrix} 0.065 & 0.058 & 0.79 & 0.12 \end{bmatrix}.$$

The random variables $w(i,j)$ and $v(i,j)$ are uncorrelated zero mean Gaussian white noises with respective covariances $Q(i,j) = 0.0125$ and $R(i,j) = 0.0016$. The measurement degradation coefficient $\beta(i,j)$ has the following probability mass function:

$$p_s(i,j) = \begin{cases} 0.05, & s = 0 \\ 0.10, & s = 0.5 \\ 0.85, & s = 1 \end{cases}$$

where the corresponding expectation and variance can be computed as $\bar{\beta}(i,j) = 0.9$ and $\tilde{\beta}(i,j) = 0.065$. The stochastic nonlinear function $f(x(i,j), \xi(i,j))$ is selected as follows:

$$f(x(i,j), \xi(i,j))$$
$$= \begin{bmatrix} 0.01 & 0.02 & 0.03 & 0.015 \end{bmatrix}^T$$
$$\times \big[0.1\text{sign}(x_1(i,j))x_1(i,j)\xi_1(i,j) + 0.2\text{sign}(x_2(i,j))x_2(i,j)\xi_2(i,j)$$
$$+ 0.1\text{sign}(x_3(i,j))x_3(i,j)\xi_3(i,j) + 0.3\text{sign}(x_4(i,j))x_4(i,j)\xi_4(i,j)\big],$$

where $x_k(i,j)$ and $\xi_k(i,j)$ ($k \in [1 \; 4]$) stand for, respectively, the k-th element of state $x(i,j)$ and $\xi(i,j)$, in which $\xi(i,j)$ is the Gaussian white noise with zero mean and variance I. It is evident to see that such a nonlinear function meets conditions (2.2b) and (2.2c) with $h = 1$ and matrices $\Pi_s = 10^{-4}[1\;2\;3\;1.5]^T[1\;2\;3\;1.5]$ and $\Gamma_s = 10^{-2}\text{diag}\{1,4,1,9\}$. Moreover, the initial conditions are chosen as $u_1(i) = u_2(j) = 0$ for $i,j \in [0 \; N]$, $P(i,0) = 0.11I$, $P(0,j) = 0.1I$ for $i,j \in [1 \; N]$, and $P(0,0) = 0.105I$.

According to the established algorithm, the filter gains and the error variances can be recursively calculated. For space consideration, only part of the results are presented as follows:

$$K(1,1) = \begin{bmatrix} 0.0862 & 0.0772 & 1.2336 & 0.1639 \end{bmatrix}^T,$$

$$K(1,2) = \begin{bmatrix} 0.4009 & 0.3592 & 0.6827 & 0.0237 \end{bmatrix}^T,$$

$$K(2,1) = \begin{bmatrix} 0.0018 & 0.0665 & 1.2404 & 0.1647 \end{bmatrix}^T,$$

$$P_u(1,1) = \begin{bmatrix} 0.0995 & -0.0004 & -0.0072 & -0.0010 \\ -0.0004 & 0.0996 & -0.0064 & -0.0008 \\ -0.0072 & -0.0064 & 0.0145 & -0.0137 \\ -0.0010 & -0.0008 & -0.0137 & 0.1010 \end{bmatrix},$$

$$P_u(1,2) = \begin{bmatrix} 0.0094 & -0.0141 & 0.0002 & -0.0004 \\ -0.0141 & 0.0944 & -0.0056 & 0.0026 \\ 0.0002 & -0.0056 & 0.0090 & -0.0142 \\ -0.0004 & 0.0026 & -0.0142 & 0.0944 \end{bmatrix},$$

$$P_u(2,1) = \begin{bmatrix} 0.0145 & -0.0137 & -0.0002 & 0 \\ -0.0137 & 0.1008 & -0.0055 & -0.0007 \\ -0.0002 & -0.0055 & 0.0139 & -0.0138 \\ 0 & -0.0007 & -0.0138 & 0.1010 \end{bmatrix}.$$

By resorting to the Monte Carlo method, the estimation error and the trace of filtering error variance are averaged in 100 independent experiments. The simulation results are shown in Figs. 2.1–2.5, where Figs. 2.1–2.4 depict the filtering error evolution $\tilde{x}_u(i,j)$ with its k-th element denoted by $\tilde{x}_u^{(k)}(i,j)$ ($k \in [1\ 4]$), and Fig. 2.5 shows the evolution of the minimal variance $P_u(i,j)$.

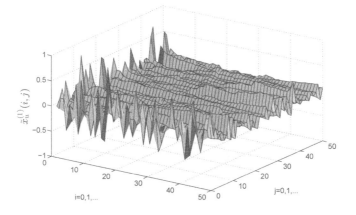

FIGURE 2.1: Estimation error $\tilde{x}_u^{(1)}(i,j)$.

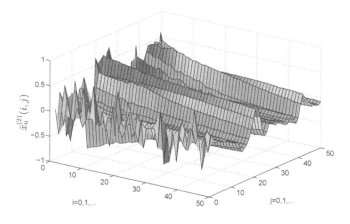

FIGURE 2.2: Estimation error $\tilde{x}_u^{(2)}(i,j)$.

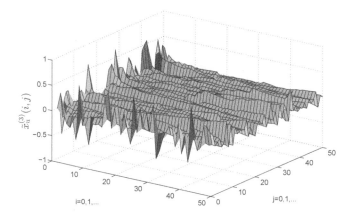

FIGURE 2.3: Estimation error $\tilde{x}_u^{(3)}(i,j)$.

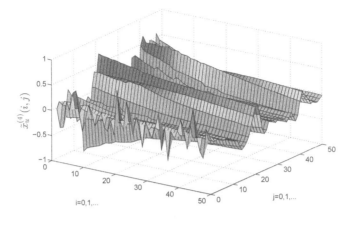

FIGURE 2.4: Estimation error $\tilde{x}_u^{(4)}(i,j)$.

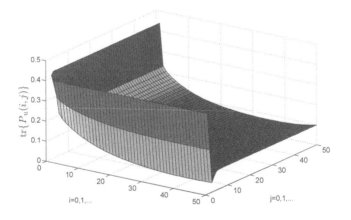

FIGURE 2.5: Trace trajectory of $P_u(i,j)$.

To reflect the characteristic that the filtering performance $\mathrm{tr}\{P_u(i,j)\}$ is related to the probability $\bar{\beta}(i,j)$ of the measurement degradation, we compare the traces of the minimal filtering error variances by choosing different values of $\bar{\beta}(i,j)$ with the same variance $\tilde{\beta}(i,j) = 0.065$. Apart from $\bar{\beta}(i,j) = 0.9$ given earlier, we also take $\bar{\beta}(i,j)$ to be 0.7 and 0.5, and denote the corresponding minimal variances as $P_u^{(1)}(i,j)$ and $P_u^{(2)}(i,j)$, respectively. The simulation results are presented in Figs. 2.6–2.7. Obviously, the trace of the minimal filtering error variance $\mathrm{tr}\{P_u(i,j)\}$ increases as $\bar{\beta}(i,j)$ declines, which is consistent with the theoretical analysis results.

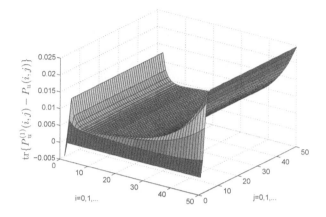

FIGURE 2.6: Difference between the traces of $P_u(i,j)$ and $P_u^{(1)}(i,j)$.

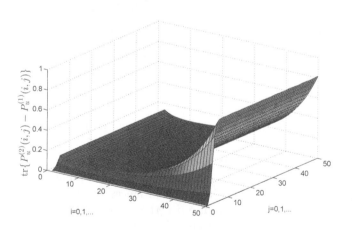

FIGURE 2.7: Difference between the traces of $P_u^{(1)}(i,j)$ and $P_u^{(2)}(i,j)$.

2.5 Summary

In this chapter, the recursive filtering problem has been investigated for 2-D shift-varying systems subject to both stochastic nonlinearity and degraded measurements over a finite horizon. The nonlinear perturbation is described via its statistical properties and the occurrence of the degraded measurements is characterized by random variables taking values on the interval $[0, 1]$. An unbiased filter has been put forward in a similar form as the 1-D Kalman filter. The desired filter has then been designed, through a set of recursive equations, in order to ensure the minimum of the filtering error variance. Moreover, not only the boundedness but also the monotonicity (with respect to the measurement degradation) has been analyzed for the filter performance. An easy-to-implement filtering algorithm has been proposed finally which can be iteratively calculated online. Simulation results have also been given to demonstrate the applicability of the proposed algorithm through an example of practical insight.

3

Robust Kalman Filtering for 2-D Systems with Multiplicative Noises and Measurement Degradations

---

It is noteworthy that the renowned Kalman filter requires accurate knowledge of the system parameters, the linear structure of the underlying system, and the exact statistics of the Gaussian noises. Such a requirement tends to be stringent since systems to be tackled may involve unmodeled dynamics caused by, for example, parameter variations. To improve the robustness of the traditional Kalman filter, the robust Kalman filter has been a recurring theme of research interests in control and signal processing communities.

For 2-D uncertain systems, the corresponding robust Kalman filtering problem has not been adequately studied yet due probably to the mathematical difficulties. One of the primary challenges is to establish a novel approach for designing the 2-D robust Kalman filter that focuses on developing the tightest upper bound (in the sense of trace) on the filtering error variance at each step. Furthermore, the bidirectional evolution leads to the difficulty in formulating the recursion of the error variances in the 2-D space. It is also a complicated issue to find the tightest bound of the error variances through appropriately designing the filter gains in the 2-D cases.

Notice that sensors are prone to experience possible failures resulting in the phenomenon of measurement degradation. For decades, a surge of research attention has been paid to the filtering problems with measurement degradation, and much progress has been made on the filter design issue of 2-D shift-invariant systems with steady-state dynamic behaviors. From the engineering perspective, however, it is often the case that the transient behaviors are practically more important than the asymptotic ones, which is particularly true when the system parameters are inherently shift-varying. As such, it is of great importance to address the robust filtering problem for 2-D shift-varying systems with guaranteed error variance as the performance index.

Stimulated by the above discussions, this chapter aims to study the robust Kalman filter design problem for 2-D shift-varying uncertain systems with measurement degradation. The general FM-II model is considered to depict the system whose states are to be estimated. Parameter uncertainties are presented in the form of unknown but norm-bounded matrices. The phenomenon of measurement degradation is characterized by random variable satisfying certain given probabilistic distribution. The main contributions of this chapter can be highlighted as follows: (1) the problem to be tackled is new and our attempt represents one of the first few to investigate the robust recursive filtering problem for 2-D systems over a finite horizon; (2) a novel approach is developed to guarantee the existence of an upper bound for the generalized estimation error variance; and (3) the robust Kalman filter is designed which ensures the local minimum trace of the acquired upper bound.

The rest of this chapter is organized as follows. Section 3.1 formulates the robust Kalman filtering problem for the considered 2-D system. Section 3.2 provides an upper bound for the generalized estimation error variance, and the determination of the locally minimal one is presented in Section 3.3. A simulation example is given to illustrate effectiveness of the theoretical results in Section 3.4. Finally, the conclusion is drawn in Section 3.5.

3.1 Problem Formulation and Preliminaries

Consider the following 2-D shift-varying uncertain system described by the FM-II state-space model over a finite horizon $i, j \in [1 \ K]$:

$$
\begin{aligned}
x(i,j) = & \left[A_1(i,j-1) + \Delta A_1(i,j-1) + \breve{A}_1(i,j-1)\xi(i,j-1) \right] x(i,j-1) \\
& + \left[A_2(i-1,j) + \Delta A_2(i-1,j) + \breve{A}_2(i-1,j)\xi(i-1,j) \right] x(i-1,j) \\
& + B_1(i,j-1)w(i,j-1) + B_2(i-1,j)w(i-1,j), \quad\quad (3.1a) \\
y(i,j) = & \left[\beta(i,j)C(i,j) + \Delta C(i,j) + D(i,j)\tau(i,j) \right] x(i,j) + v(i,j), \quad (3.1b)
\end{aligned}
$$

where $x(i,j) \in \mathbb{R}^n$ and $y(i,j) \in \mathbb{R}^m$ are the state and the measured output, respectively, $\xi(i,j)$ and $\tau(i,j) \in \mathbb{R}$ are the multiplicative noises with mean zero and respective variances σ_ξ^2 and σ_τ^2, $w(i,j) \in \mathbb{R}^p$ is the additive noise and $v(i,j) \in \mathbb{R}^m$ is the measurement noise which are both white sequences with zero means. The matrices $A_l(i,j)$, $\breve{A}_l(i,j)$, $B_l(i,j)$ $(l = 1,2)$, $C(i,j)$, and $D(i,j)$ describe the nominal system, while the uncertain matrices $\Delta A_l(i,j)$ $(l = 1,2)$ and $\Delta C(i,j)$ represent the parameter perturbations having the following structure:

$$
\begin{aligned}
\Delta A_l(i,j) &\triangleq N_l(i,j)F(i,j)M(i,j), \\
\Delta C(i,j) &\triangleq E(i,j)F(i,j)M(i,j),
\end{aligned}
$$

where $N_l(i,j)$, $E(i,j)$, and $M(i,j)$ are known matrices with appropriate dimensions, and $F(i,j)$ is an uncertain one with $F(i,j)F^T(i,j) \leq I$. The stochastic variable $\beta(i,j) \in \mathbb{R}$ governing the measurement degradation distributes over the interval $[0,1]$ with $\mathbb{E}\{\beta(i,j)\} = \bar{\beta}(i,j)$ and $\text{Var}\{\beta(i,j)\} = \hat{\beta}(i,j)$, where $\bar{\beta}(i,j)$ and $\hat{\beta}(i,j)$ are known constants. The initial boundary conditions $x(i,0)$ and $x(0,j)$ associated with system (3.1) satisfy $\mathbb{E}\{x(i,0)\} = u_1(i)$, $\mathbb{E}\{x(0,j)\} = u_2(j)$, $\text{Cov}\{x(i,0), x(i,0)\} = \Xi(i,0)$, and $\text{Cov}\{x(0,j), x(0,j)\} = \Xi(0,j)$ for $i, j \in [1\ K]$, where $u_1(i)$ and $u_2(j)$ are given vectors, $\Xi(i,0)$ and $\Xi(0,j) > 0$ are known positive definite matrices.

The following assumptions are made throughout this chapter.

Assumption 3.1 *For $i, j, k, l \in [0\ K]$, the following correlation is satisfied*

$$\mathbb{E}\{\eta(i,j)\eta^T(k,l)\} = \text{diag}\{\sigma_\xi^2, \hat{\beta}(i,j), \sigma_\tau^2, Q(i,j), R(i,j)\}\delta(i,k)\delta(j,l)$$

where

$$\eta(i,j) \triangleq [\xi(i,j)\ \ \tilde{\beta}(i,j)\ \ \tau(i,j)\ \ w^T(i,j)\ \ v^T(i,j)]^T,$$
$$\tilde{\beta}(i,j) \triangleq \beta(i,j) - \bar{\beta}(i,j),$$

$\delta(i,k)$ *is the Kronecker delta function with $\delta(i,k)$ being unity for $i = k$ and zero otherwise, $Q(i,j) \geq 0$ and $R(i,j) > 0$ are known matrices signifying the statistics of the second moments for the noises.*

Assumption 3.2 *For $i, j \in [0\ K]$, the initial states $x(i,0)$ and $x(0,j)$ are independent of all the other random variables involved in system (3.1).*

In this chapter, the following filter is adopted for system (3.1):

$$\hat{x}(i,j) = \hat{A}_1(i,j-1)\hat{x}(i,j-1) + \hat{A}_2(i-1,j)\hat{x}(i-1,j)$$
$$+ \hat{K}_1(i,j-1)\big(y(i,j-1) - \bar{\beta}(i,j-1)C(i,j-1)\hat{x}(i,j-1)\big)$$
$$+ \hat{K}_2(i-1,j)\big(y(i-1,j) - \bar{\beta}(i-1,j)C(i-1,j)\hat{x}(i-1,j)\big), \quad (3.2)$$

where $\hat{x}(i,j) \in \mathbb{R}^n$ is the state estimate, and $\hat{A}_l(i,j)$ and $\hat{K}_l(i,j)$ $(l = 1, 2)$ are filter parameters to be determined. The initial condition related to filter (3.2) is given as $\hat{x}(i,0) = \hat{x}(0,j) = 0$ for $i, j \in [1\ K]$.

Remark 3.1 *Filter (3.2) updates the state estimate based on the innovation $y(i,j) - \bar{\beta}(i,j)C(i,j)\hat{x}(i,j)$ along two independent directions, which presents a similar form as the traditional Kalman filter for 1-D models. This filter is recognized as a 2-D robust Kalman filter, which is constructed for the convenience of establishing the locally optimal upper bound for the second-order moment of the estimation error.*

The objective of this chapter is twofold. First, we aim to design a filter of form (3.2) and find a sequence of positive definite matrices $\{S(i,j)|i,j \in [0\ K]\}$ such that the second-order moment of the estimation error is bounded by

$$\tilde{\Xi}(i,j) \triangleq \mathbb{E}\left\{[x(i,j) - \hat{x}(i,j)][x(i,j) - \hat{x}(i,j)]^T\right\} \leq S(i,j) \quad (3.3)$$

where $\tilde{\Xi}(i,j)$ is viewed as the generalized estimation error variance in a broad sense throughout this chapter. Then, we intend to locally minimize such a bound $S(i,j)$ at each shift step in the trace sense by properly determining the filter parameters. The problem to be addressed is referred to as *the robust Kalman filtering problem for 2-D shift-varying system over a finite horizon.*

Remark 3.2 *In order to meet the practical requirements with an acceptable filtering performance, the system states are expected to be confined in a moving "window" whose size is governed by the upper bound of the estimation error variance. In this chapter, such a filtering performance constraint is described naturally by the locally minimal upper bound (rather than the global one) of the error variance at each shift step. Furthermore, for shift-varying systems, evolution of the 2-D robust Kalman filter has a similar execution feature as that of the traditional 1-D Kalman filter, that is, filter at the current coordinate (i,j) will contribute to the computation of filter parameters at "future" coordinates (i.e., coordinates with indexes larger than the current one with indexes i and j). Such an identity of robust Kalman filter is induced inevitably by its recursive form and transient feature, where the desired filter parameters can be calculated iteratively.*

To solve the addressed robust Kalman filtering problem, the following lemmas are introduced which will be used in the later analysis.

Lemma 3.1 *Let A and B be real-valued matrices with appropriate dimensions. For any given positive scalar $\mu > 0$, the following inequality holds*

$$AB^T + BA^T \leq \mu AA^T + \mu^{-1}BB^T.$$

Lemma 3.2 *[206] Assume that A, N, M, and F are matrices with compatible dimensions and F satisfies $FF^T \leq I$. Let X and Y be matrices satisfying $Y \geq X > 0$, and ϵ be an arbitrary positive constant satisfying $\epsilon^{-1}I - MYM^T > 0$. Suppose that $f(\cdot)$ and $g(\cdot)$ are matrix functions defined as follows:*

$$f(X) \triangleq (A + NFM)X(A + NFM)^T,$$
$$g(X) \triangleq A(X^{-1} - \epsilon M^T M)^{-1}A^T + \epsilon^{-1}NN^T.$$

Then, the following inequalities hold:

$$f(X) \leq f(Y), \quad g(X) \leq g(Y), \quad f(X) \leq g(X).$$

Lemma 3.3 *Assume that A_l, N_l ($l = 1,2$), M, and F are matrices with compatible dimensions and F satisfies $FF^T \leq I$. Let $\epsilon_l(i,j) > 0$ be arbitrary positive constants satisfying $\epsilon_l^{-1}(i,j)I - M\bar{Z}(i,j)M^T > 0$, where $\bar{Z}(i,j)$ are positive definite matrices satisfying $\bar{Z}(i,j) \geq Z(i,j) \triangleq \mathbb{E}\{z(i,j)z^T(i,j)\}$ with $z(i,j)$ ($i,j \in [0\ K]$) being 2-D state vectors. For any given positive scalar $\mu > 0$, suppose that $\bar{f}(\cdot,\cdot)$ and $\bar{g}(\cdot,\cdot)$ are matrix functions defined as follows:*

$$\bar{f}(z(i,j-1), z(i-1,j))$$

$$\triangleq \mathbb{E}\left\{(A_1 + N_1 FM)z(i,j-1)z^T(i-1,j)(A_2 + N_2 FM)^T\right\}$$
$$+ \mathbb{E}\left\{(A_2 + N_2 FM)z(i-1,j)z^T(i,j-1)(A_1 + N_1 FM)^T\right\}, \tag{3.4}$$

$$\bar{g}(\bar{Z}(i,j-1), \bar{Z}(i-1,j))$$
$$\triangleq \mu\big[A_1(\bar{Z}^{-1}(i,j-1) - \epsilon_1(i,j-1)M^T M)^{-1}A_1^T$$
$$+ \epsilon_1^{-1}(i,j-1)N_1 N_1^T\big] + \mu^{-1}\big[A_2(\bar{Z}^{-1}(i-1,j)$$
$$- \epsilon_2(i-1,j)M^T M)^{-1}A_2^T + \epsilon_2^{-1}(i-1,j)N_2 N_2^T\big]. \tag{3.5}$$

Then, the following inequality holds for $i,j \in [1\ K]$:

$$\bar{f}(z(i,j-1), z(i-1,j)) \le \bar{g}(\bar{Z}(i,j-1), \bar{Z}(i-1,j)). \tag{3.6}$$

Proof *By resorting to Lemmas 3.1 and 3.2, one has*

$$\bar{f}(z(i,j-1), z(i-1,j))$$
$$\le \mu(A_1 + N_1 FM)Z(i,j-1)(A_1 + N_1 FM)^T$$
$$+ \mu^{-1}(A_2 + N_2 FM)Z(i-1,j)(A_2 + N_2 FM)^T$$
$$\le \mu(A_1 + N_1 FM)\bar{Z}(i,j-1)(A_1 + N_1 FM)^T$$
$$+ \mu^{-1}(A_2 + N_2 FM)\bar{Z}(i-1,j)(A_2 + N_2 FM)^T$$
$$\le \bar{g}(\bar{Z}(i,j-1), \bar{Z}(i-1,j))$$

which completes the proof.

Lemma 3.3 presents a useful approach which constructs an upper bound $\bar{g}(\cdot,\cdot)$ for the matrix function $\bar{f}(\cdot,\cdot)$ with norm-bounded uncertainties.

3.2 Upper Bound for the Generalized Error Variance

Noticing that the parameter uncertainties are involved in the 2-D system (3.1), it is impossible to accurately derive the estimation error variance. Therefore, instead of obtaining the analytical expression for the exact error variance, an alternative way is to derive an upper bound for the generalized estimation error variance and subsequently minimize it.

Define $\bar{x}(i,j) \triangleq [x^T(i,j)\ \hat{x}^T(i,j)]^T$. It follows from (3.1) and (3.2) that

$$\bar{x}(i,j) = \big[\bar{A}_1(i,j-1) + \Delta\bar{A}_1(i,j-1) + \tilde{A}_1(i,j-1)\big]\bar{x}(i,j-1)$$
$$+ \big[\bar{A}_2(i-1,j) + \Delta\bar{A}_2(i-1,j) + \tilde{A}_2(i-1,j)\big]\bar{x}(i-1,j)$$
$$+ \bar{B}_1(i,j-1)\bar{w}(i,j-1) + \bar{B}_2(i-1,j)\bar{w}(i-1,j) \tag{3.7}$$

where, for $l = 1, 2$, $\bar{B}_l(i,j) = \text{diag}\{B_l(i,j), \hat{K}_l(i,j)\}$,

$$\bar{A}_l(i,j) = \begin{bmatrix} A_l(i,j) & 0 \\ \bar{\beta}(i,j)\hat{K}_l(i,j)C(i,j) & \hat{A}_l(i,j) \end{bmatrix},$$

$$\Delta \bar{A}_l(i,j) = \bar{N}_l(i,j) F(i,j) \bar{M}(i,j),$$

$$\tilde{A}_l(i,j) = \begin{bmatrix} \xi(i,j) \breve{A}_l(i,j) & 0 \\ \acute{K}_l(i,j) & 0 \end{bmatrix}, \quad \bar{w}(i,j) = \begin{bmatrix} w(i,j) \\ v(i,j) \end{bmatrix},$$

$$\bar{N}_l(i,j) = \begin{bmatrix} N_l(i,j) \\ \hat{K}_l(i,j) E(i,j) \end{bmatrix}, \quad \bar{M}(i,j) = [M(i,j) \quad 0],$$

$$\acute{A}_l(i,j) = \hat{A}_l(i,j) - \bar{\beta}(i,j) \hat{K}_l(i,j) C(i,j),$$

$$\acute{K}_l(i,j) = \bar{\beta}(i,j) \hat{K}_l(i,j) C(i,j) + \tau(i,j) \hat{K}_l(i,j) D(i,j).$$

Recalling the statistical properties of random variables $\xi(i,j)$, $\tau(i,j)$, and $\beta(i,j)$, one can verify that

$$\Pi_l(i,j) \triangleq \mathbb{E}\{\tilde{A}_l(i,j) \bar{x}(i,j) \bar{x}^T(i,j) \tilde{A}_l^T(i,j)\}$$
$$= \operatorname{diag}\{\sigma_\xi^2 \breve{A}_l(i,j) \Xi(i,j) \breve{A}_l^T(i,j), \bar{\Xi}_l(i,j)\} \tag{3.8}$$

where $\Xi(i,j) \triangleq \mathbb{E}\{x(i,j) x^T(i,j)\}$ and

$$\bar{\Xi}_l(i,j) \triangleq \hat{\beta}(i,j) \hat{K}_l(i,j) C(i,j) \Xi(i,j) C^T(i,j) \hat{K}_l^T(i,j)$$
$$+ \sigma_\tau^2 \hat{K}_l(i,j) D(i,j) \Xi(i,j) D^T(i,j) \hat{K}_l^T(i,j).$$

Then, according to (3.7) and (3.8), the dynamic evolution of $\bar{\Theta}(i,j) \triangleq \mathbb{E}\{\bar{x}(i,j) \bar{x}^T(i,j)\}$ is obtained as follows:

$$\bar{\Theta}(i,j) = f_1((i,j-1), \bar{\Theta}(i,j-1)) + f_2((i-1,j), \bar{\Theta}(i-1,j))$$
$$+ f_3(\bar{x}(i,j-1), \bar{x}(i-1,j)) \tag{3.9}$$

where, for the given matrix X and $l = 1, 2$,

$$f_l((i,j), X) \triangleq \Pi_l(i,j) + [\bar{A}_l(i,j) + \Delta \bar{A}_l(i,j)] X [\bar{A}_l(i,j) + \Delta \bar{A}_l(i,j)]^T$$
$$+ \bar{B}_l(i,j) \bar{Q}(i,j) \bar{B}_l^T(i,j),$$

$$f_3(\bar{x}(i,j-1), \bar{x}(i-1,j)) \triangleq \Phi(\bar{x}(i,j-1), \bar{x}(i-1,j))$$
$$+ \Phi^T(\bar{x}(i,j-1), \bar{x}(i-1,j))$$

in which $\bar{Q}(i,j) \triangleq \operatorname{diag}\{Q(i,j), R(i,j)\}$ and

$$\Phi(\bar{x}(i,j-1), \bar{x}(i-1,j))$$
$$\triangleq \mathbb{E}\{[\bar{A}_1(i,j-1) + \Delta \bar{A}_1(i,j-1)] \bar{x}(i,j-1) \bar{x}^T(i-1,j)$$
$$\times [\bar{A}_2(i-1,j) + \Delta \bar{A}_2(i-1,j)]^T\}.$$

In [244], a concept of 1-D identity quadratic filter has been introduced, as the extension of the classical Kalman filter, in order to obtain the upper bound of the error estimation variance for systems with parameter uncertainties. Along the similar line as in [244], we consider the following definition for 2-D system (3.1).

Definition 3.1 *The filter (3.2) is said to be a 2-D identity quadratic filter associated with $\Theta(i,j)$ and $P(i,j)$ for system (3.1) if there are two sequences of positive definite matrices $\{P(i,j)\}_{i,j=0}^{K}$ and $\{\Theta(i,j)\}_{i,j=0}^{K}$ as well as two sequences of positive scalars $\{\epsilon_l(i,j)\}_{i,j=0}^{K}$ $(l=1,2)$ such that the following relationships*

$$\epsilon_l^{-1}(i,j)I - \bar{M}(i,j)\Theta(i,j)\bar{M}^T(i,j) > 0, \tag{3.10}$$

$$\Theta(i,j) = h_1((i,j-1),\Theta(i,j-1)) + h_2((i-1,j),\Theta(i-1,j))$$
$$+ h_3(\Theta(i,j-1),\Theta(i-1,j)) \tag{3.11}$$

are satisfied, where, for a given matrix X and constant $\mu > 0$,

$$h_l((i,j),X) \triangleq g_l((i,j),X) + \bar{\Pi}_l(i,j) + \bar{B}_l(i,j)\bar{Q}(i,j)\bar{B}_l^T(i,j),$$

$$g_l((i,j),X) \triangleq \bar{A}_l(i,j)\left[X^{-1} - \epsilon_l(i,j)\bar{M}^T(i,j)\bar{M}(i,j)\right]^{-1}\bar{A}_l^T(i,j)$$
$$+ \epsilon_l^{-1}(i,j)\bar{N}_l(i,j)\bar{N}_l^T(i,j),$$

$$\bar{\Pi}_l(i,j) \triangleq \text{diag}\{\sigma_\xi^2 \breve{A}_l(i,j)P(i,j)\breve{A}_l^T(i,j), \tilde{K}_l(i,j)\},$$

$$\tilde{K}_l(i,j) \triangleq \hat{\beta}(i,j)\hat{K}_l(i,j)C(i,j)P(i,j)C^T(i,j)\hat{K}_l^T(i,j)$$
$$+ \sigma_\tau^2 \hat{K}_l(i,j)D(i,j)P(i,j)D^T(i,j)\hat{K}_l^T(i,j),$$

$$h_3(\Theta(i,j-1),\Theta(i-1,j)) \triangleq \mu^{-1}g_2((i-1,j),\Theta(i-1,j))$$
$$+ \mu g_1((i,j-1),\Theta(i,j-1)).$$

The following theorem is provided to show the existence of an upper bound for matrix $\bar{\Theta}(i,j)$.

Theorem 3.1 *Assume that (3.2) is a 2-D identity quadratic filter associated with $\Theta(i,j)$ and $P(i,j)$ for system (3.1). For $i,j \in [0\ K]$, if $\Xi(i,j) \leq P(i,j)$, $\bar{\Theta}(i,0) \leq \Theta(i,0)$, and $\bar{\Theta}(0,j) \leq \Theta(0,j)$, then matrix $\bar{\Theta}(i,j)$ is bounded by*

$$\bar{\Theta}(i,j) \leq \Theta(i,j). \tag{3.12}$$

Proof *To begin with, one has $\Pi_l(i,j) \leq \bar{\Pi}_l(i,j)$ by resorting to the condition $\Xi(i,j) \leq P(i,j)$. It can be checked that, for any given matrix X with $X \leq \Theta(i,j)$ and $l = 1,2$, the following relationships are satisfied:*

$$f_l((i,j),X) \leq f_l((i,j),\Theta(i,j)) \leq h_l((i,j),\Theta(i,j)) \tag{3.13}$$

where condition (3.10) and Lemma 3.2 have been utilized in the last step.

The assertion of (3.12) is provided by induction on the indices i and j. In the following, we will first analyze the relationship between $\bar{\Theta}(i,1)$ and $\Theta(i,1)$ as well as the relationship between $\bar{\Theta}(1,j)$ and $\Theta(1,j)$, respectively. It follows from (3.9), (3.11), and (3.13) that

$$\bar{\Theta}(1,1) \leq f_1((1,0),\Theta(1,0)) + f_2((0,1),\Theta(0,1)) + h_3(\Theta(1,0),\Theta(0,1))$$
$$\leq \Theta(1,1) \tag{3.14}$$

where Lemma 3.3 has been utilized in the above derivation.

Assume that inequality $\bar{\Theta}(k,1) \leq \Theta(k,1)$ is true for a given integer $k \in [1 \quad K-1]$. Then, one has

$$\bar{\Theta}(k+1,1) \leq f_1((k+1,0), \Theta(k+1,0)) + f_2((k,1), \Theta(k,1))$$
$$+ h_3(\Theta(k+1,0), \Theta(k,1))$$
$$\leq \Theta(k+1,1). \tag{3.15}$$

Similarly, with the validity of $\bar{\Theta}(1,l) \leq \Theta(1,l)$ for a given integer $l \in [1 \quad K-1]$, the following inequality is satisfied:

$$\bar{\Theta}(1,l+1) \leq f_1((1,l), \Theta(1,l)) + f_2((0,l+1), \Theta(0,l+1))$$
$$+ h_3(\Theta(1,l), \Theta(0,l+1))$$
$$\leq \Theta(1,l+1). \tag{3.16}$$

Based on the inductive approach, we have $\bar{\Theta}(i,1) \leq \Theta(i,1)$ and $\bar{\Theta}(1,j) \leq \Theta(1,j)$ for all $i,j \in [0\ K]$. In other words, the inequality $\bar{\Theta}(i,j) \leq \Theta(i,j)$ is satisfied for $(i,j) \in \{(i_0,1)|1 \leq i_0 \leq K\} \cup \{(1,j_0)|1 < j_0 \leq K\}$.

For further development, we assume that $\bar{\Theta}(i,j) \leq \Theta(i,j)$ is true for $(i,j) \in \{(i_0,l)|k \leq i_0 \leq K\} \cup \{(k,j_0)|l < j_0 \leq K\}$, where $k,l \in [1 \quad K-1]$ are given constants. Then, the following inequality is derived:

$$\bar{\Theta}(k+1,l+1) \leq f_1((k+1,l), \Theta(k+1,l)) + f_2((k,l+1), \Theta(k,l+1))$$
$$+ h_3(\Theta(k+1,l), \Theta(k,l+1))$$
$$\leq \Theta(k+1,l+1).$$

Adopting a similar procedure as in the derivation of (3.15) and (3.16), one has

$$\bar{\Theta}(i,l+1) \leq \Theta(i,l+1), \quad i \in [k+1 \quad K]$$
$$\bar{\Theta}(k+1,j) \leq \Theta(k+1,j), \quad j \in [l+1 \quad K]$$

which means $\bar{\Theta}(i,j) \leq \Theta(i,j)$ holds for all $(i,j) \in \{(i_0,l+1)|k+1 \leq i_0 \leq K\} \cup \{(k+1,j_0)|l+1 < j_0 \leq K\}$. Repeating the above process iteratively with regard to i and j line-by-line, we arrive at $\bar{\Theta}(i,j) \leq \Theta(i,j)$ for all $i,j \in [0\ K]$, which completes the proof.

According to Theorem 3.1, the following corollary is obtained directly.

Corollary 3.1 *Under the same conditions as in Theorem 3.1, the following inequality is satisfied:*

$$\tilde{\Xi}(i,j) \leq [I \ - I]\Theta(i,j)[I \ - I]^T, \quad i,j \in [1\ K]. \tag{3.17}$$

From Theorem 3.1, an upper bound $\Theta(i,j)$ for matrix $\bar{\Theta}(i,j)$ is derived subject to $\Xi(i,j) \leq P(i,j)$. Therefore, a matrix $P(i,j)$ should be provided that satisfies $\Xi(i,j) \leq P(i,j)$ before constructing $\Theta(i,j)$. In addition, the solution to the Riccati-like difference equation (3.11) may not be unique which, according to Corollary 3.1, would further result in the non-uniqueness of the upper bound for the generalized estimation error variance $\tilde{\Xi}(i,j)$. Hence, it is essential to minimize the upper bound in the sense of trace for $\tilde{\Xi}(i,j)$ by properly selecting filter parameters $\hat{A}_l(i,j)$ and $\hat{K}_l(i,j)$ $(l=1,2)$.

3.3 Suboptimal Filter Design

This section presents a sufficient condition ensuring the existence of the 2-D identity quadratic filter for system (3.1), which would help obtain an upper bound for the generalized estimation error variance. In addition, a suboptimal filter is designed by minimizing the obtained bound in the trace sense.

To begin with, the following lemma is introduced.

Lemma 3.4 *For $i,j \in [0\ K]$, suppose that $\tilde{f}_l((i,j),\cdot)$ and $\tilde{g}_l((i,j),\cdot)$ $(l=1,2)$ are symmetric matrix functions satisfying*

$$\tilde{f}_l((i,j),X) < \tilde{g}_l((i,j),Y)$$

with $X < Y$. Then, the solutions $A(i,j)$ and $B(i,j)$ (with initial conditions $A(i,0) < B(i,0)$ and $A(0,j) < B(0,j)$) of the following recursive equations

$$A(i,j) = \tilde{f}_1((i,j-1),A(i,j-1)) + \tilde{f}_2((i-1,j),A(i-1,j)), \qquad (3.18)$$
$$B(i,j) = \tilde{g}_1((i,j-1),B(i,j-1)) + \tilde{g}_2((i-1,j),B(i-1,j)) \qquad (3.19)$$

satisfy the inequality

$$A(i,j) < B(i,j). \qquad (3.20)$$

Proof *Along the similar line in the proof of Theorem 3.1, we now prove Lemma 3.4 by induction. In the light of (3.18), (3.19), and the initial conditions, it is easy to derive that*

$$
\begin{aligned}
A(1,1) &= \tilde{f}_1((1,0),A(1,0)) + \tilde{f}_2((0,1),A(0,1)) \\
&< \tilde{g}_1((1,0),B(1,0)) + \tilde{g}_2((0,1),B(0,1)) \\
&= B(1,1).
\end{aligned}
$$

Assuming that $A(k,1) < B(k,1)$ and $A(1,l) < B(1,l)$ are true for given integers $k,l \in [1\ K-1]$, we obtain

$$
\begin{aligned}
A(k+1,1) &= \tilde{f}_1((k+1,0),A(k+1,0)) + \tilde{f}_2((k,1),A(k,1)) \\
&< \tilde{g}_1((k+1,0),B(k+1,0)) + \tilde{g}_2((k,1),B(k,1)), \\
&= B(k+1,1)
\end{aligned}
$$

$$A(1, l+1) = \tilde{f}_1((1, l), A(1, l)) + \tilde{f}_2((0, l+1), A(0, l+1))$$
$$< \tilde{g}_1((1, l), B(1, l)) + \tilde{g}_2((0, l+1), B(0, l+1))$$
$$= B(1, l+1).$$

Therefore, inequality (3.20) is true for all $(i, j) \in \{(i_0, 1)|1 \leq i_0 \leq K\} \cup \{(1, j_0)|1 < j_0 \leq K\}$ *by mathematical induction.*

For given integers $k, l \in [1 \quad K-1]$, *further assume that (3.20) is valid for all* $(i, j) \in \{(i_0, l)|k \leq i_0 \leq K\} \cup \{(k, j_0)|l < j_0 \leq K\}$. *Then, it is clear to see*

$$A(k+1, l+1) = \tilde{f}_1((k+1, l), A(k+1, l)) + \tilde{f}_2((k, l+1), A(k, l+1))$$
$$< \tilde{g}_1((k+1, l), B(k+1, l)) + \tilde{g}_2((k, l+1), B(k, l+1))$$
$$= B(k+1, l+1)$$

which infers $A(i, j) \leq B(i, j)$ *for* $(i, j) \in \{(i_0, l+1)|k+1 \leq i_0 \leq K\} \cup \{(k+1, j_0)|l+1 < j_0 \leq K\}$ *from the inductive method. By repeating the above procedure iteratively, we conclude that (3.20) holds for all* $i, j \in [0 \ K]$, *which ends the proof.*

Remark 3.3 *Lemma 3.4 is viewed as the 2-D version of the comparison-like principle. Note that the corresponding comparison-like principle for 1-D discrete-time systems has been introduced in [190] and then widely utilized in the literature, see e.g., [214, 243]. In the sequel, the extended 2-D comparison-like principle will be exploited to ensure the existence of the 2-D identity quadratic filter.*

Let X, Y be positive definite matrices, and μ, $\epsilon_l(i, j)$ $(l = 1, 2)$ be given positive constants. For presentation simplicity, some notations associated with system (3.1) are defined as follows:

$$\mu_1 \triangleq \mu, \quad \mu_2 \triangleq \mu^{-1},$$

$$\Omega_l((i, j), X) \triangleq [X^{-1} - \epsilon_l(i, j)M^T(i, j)M(i, j)]^{-1},$$

$$\bar{\Omega}_l((i, j), X) \triangleq [\epsilon_l^{-1}(i, j)I - M(i, j)XM^T(i, j)]^{-1},$$

$$\Lambda_l((i, j), X) \triangleq \sigma_\xi^2 \breve{A}_l(i, j)X\breve{A}_l^T(i, j) + B_l(i, j)Q(i, j)B_l^T(i, j),$$

$$\bar{g}_l((i, j), X) \triangleq A_l(i, j)\Omega_l((i, j), X)A_l^T(i, j) + \epsilon_l^{-1}(i, j)N_l(i, j)N_l^T(i, j),$$

$$\Upsilon_l((i, j), X, Y) \triangleq (1 + \mu_l)[\bar{\beta}^2(i, j)C(i, j)\Omega_l((i, j), X)C^T(i, j)$$
$$+ \epsilon_l^{-1}E(i, j)E^T(i, j)] + R(i, j)$$
$$+ \hat{\beta}(i, j)C(i, j)YC^T(i, j) + \sigma_\tau^2 D(i, j)YD^T(i, j),$$

$$\tilde{\Omega}_l((i, j), X, Y) \triangleq (1 + \mu_l)^2[\bar{\beta}(i, j)A_l(i, j)\Omega_l((i, j), X)C^T(i, j)$$
$$+ \epsilon_l^{-1}N_l(i, j)E^T(i, j)]\Upsilon_l^{-1}((i, j), X, Y)$$
$$\times [\bar{\beta}(i, j)C(i, j)\Omega_l((i, j), X)A_l^T(i, j) + \epsilon_l^{-1}E(i, j)N_l^T(i, j)],$$

$$\bar{h}_l((i, j), X) \triangleq \bar{g}_l((i, j), X) + \Lambda_l((i, j), X),$$

$$\bar{h}_3(P(i,j-1), P(i-1,j)) \triangleq \mu_1 \bar{g}_1((i,j-1), P(i,j-1))$$
$$+ \mu_2 \bar{g}_2((i-1,j), P(i-1,j)), \quad \forall P(i,j) > 0.$$

Before proceeding further, the following assumption is made.

Assumption 3.3 *For $i, j \in [0\ K]$, matrices $A_l(i,j)$ with $l = 1, 2$ are invertible.*

The following theorem provides a sufficient condition under which a desired filter exists that minimizes the upper bound of the generalized estimation error variance in the trace sense.

Theorem 3.2 *For a given scalar $\mu > 0$, assume that there are two sequences of positive definite matrices $\{P(i,j)\}_{i,j=0}^{K}$ and $\{S(i,j)\}_{i,j=0}^{K}$ as well as two sequences of positive scalars $\{\epsilon_l(i,j)\}_{i,j=0}^{K}$ $(l = 1, 2)$ such that the following relationships*

$$S(i,j) = (1 + \mu_1) \big[A_1(i,j-1)\Omega_1((i,j-1), S(i,j-1))A_1^T(i,j-1)$$
$$+ \epsilon_1^{-1}(i,j-1)N_1(i,j-1)N_1^T(i,j-1) \big]$$
$$+ (1 + \mu_2) \big[A_2(i-1,j)\Omega_2((i-1,j), S(i-1,j))A_2^T(i-1,j)$$
$$+ \epsilon_2^{-1}(i-1,j)N_2(i-1,j)N_2^T(i-1,j) \big]$$
$$+ \Lambda_1((i,j-1), P(i,j-1)) + \Lambda_2((i-1,j), P(i-1,j))$$
$$- \tilde{\Omega}_1((i,j-1), S(i,j-1), P(i,j-1))$$
$$- \tilde{\Omega}_2((i-1,j), S(i-1,j), P(i-1,j)), \tag{3.21}$$
$$P(i,j) = \bar{h}_1((i,j-1), P(i,j-1)) + \bar{h}_2((i-1,j), P(i-1,j))$$
$$+ \bar{h}_3(P(i,j-1), P(i-1,j)), \tag{3.22}$$
$$\epsilon_l^{-1}(i,j)I > M(i,j)P(i,j)M^T(i,j) \tag{3.23}$$

are satisfied with initial constraints $S(i,0) = \Xi(i,0)$, $S(0,j) = \Xi(0,j)$, $P(i,0) > \Xi(i,0)$, and $P(0,j) > \Xi(0,j)$. Then, there is a 2-D identity quadratic filter for system (3.1) with parameters determined as

$$\hat{A}_l(i,j) = A_l(i,j) + \big[A_l(i,j) - \bar{\beta}(i,j)\hat{K}_l(i,j)C(i,j) \big]$$
$$\times S(i,j)M^T(i,j)\bar{\Omega}_l((i,j), S(i,j))M(i,j), \tag{3.24}$$
$$\hat{K}_l(i,j) = (1 + \mu_l) \big[\bar{\beta}(i,j)A_l(i,j)\Omega_l((i,j), S(i,j))C^T(i,j)$$
$$+ \epsilon_l^{-1}(i,j)N_l(i,j)E^T(i,j) \big] \Upsilon_l^{-1}((i,j), S(i,j), P(i,j)) \tag{3.25}$$

which ensures the generalized estimation error variance to be bounded by

$$\tilde{\Xi}(i,j) \leq S(i,j). \tag{3.26}$$

Moreover, (3.24) and (3.25) are the desired filter parameters minimizing the upper bound $S(i,j)$ in the sense of trace.

Proof *The proof is divided into three parts.*

In the first part, let us show that $P(i,j)$ is an upper bound for $\Xi(i,j)$. According to the definition of $\Xi(i,j)$, we have

$$\Xi(i,j) = \bar{f}_1((i,j-1), \Xi(i,j-1)) + \bar{f}_2((i-1,j), \Xi(i-1,j)) \\ + \bar{f}_3(x(i,j-1), x(i-1,j)) \tag{3.27}$$

where, for the given matrix X and $l = 1, 2$,

$$\bar{f}_l((i,j), X) \triangleq [A_l(i,j) + \Delta A_l(i,j)]X[A_l(i,j) + \Delta A_l(i,j)]^T + \Lambda_l((i,j), X),$$
$$\bar{f}_3(x(i,j-1), x(i-1,j)) \triangleq \Gamma(x(i,j-1), x(i-1,j)) \\ + \Gamma^T(x(i,j-1), x(i-1,j))$$

in which

$$\Gamma(x(i,j-1), x(i-1,j)) \triangleq \mathbb{E}\{[A_1(i,j-1) + \Delta A_1(i,j-1)]x(i,j-1) \\ \times x^T(i-1,j)[A_2(i-1,j) + \Delta A_2(i-1,j)]^T\}.$$

Under the condition that $Y \geq X > 0$ and $\epsilon_l^{-1}(i,j)I - M(i,j)YM^T(i,j) > 0$, it is easy to have $\epsilon_l^{-1}(i,j)I - M(i,j)XM^T(i,j) > 0$ and $\Omega_l((i,j), Y) \geq \Omega_l((i,j), X) > 0$; these combining with Lemma 3.2 infer the validity of the following inequalities:

$$\bar{f}_l((i,j), X) \leq \bar{f}_l((i,j), Y), \tag{3.28a}$$
$$\bar{h}_l((i,j), X) \leq \bar{h}_l((i,j), Y), \tag{3.28b}$$
$$\bar{f}_l((i,j), X) \leq \bar{h}_l((i,j), X), \quad l = 1, 2. \tag{3.28c}$$

From (3.22) to (3.27), it is easy to see that $P(i,j)$ and $\Xi(i,j)$ have the similar structures with those of $\Theta(i,j)$ and $\bar{\Theta}(i,j)$ presented in (3.11) and (3.9), respectively. Hence, by considering the inequality condition (3.23), the initial constraints $\Xi(i,0) < P(i,0)$ and $\Xi(0,j) < P(0,j)$, and using the same argument as in the proof of Theorem 3.1, one will derive

$$\Xi(i,j) \leq P(i,j), \quad i, j \in [0 \; K].$$

In the second part of the proof, we aim to construct a positive definite matrix $\Theta(i,j)$ satisfying (3.11). For given matrices X and Y with $Y > X > 0$ and $\epsilon_l^{-1}(i,j)I - M(i,j)YM^T(i,j) > 0$, one can get

$$0 < \Omega_l((i,j), X) < \Omega_l((i,j), Y). \tag{3.29}$$

Define matrix functions $\tilde{f}_l((i,j), X)$ and $\tilde{g}_l((i,j), Y)$ ($l = 1, 2$) as follows:

$$\tilde{f}_l((i,j), X) \triangleq \Lambda_l((i,j), Y) + (1 + \mu_l)\big[A_l(i,j)\Omega_l((i,j), X)A_l^T(i,j) \\ + \epsilon_l^{-1}(i,j)N_l(i,j)N_l^T(i,j)\big] - \tilde{\Omega}_l((i,j), X, Y), \tag{3.30}$$

$$\tilde{g}_l((i,j),Y) \triangleq (1+\mu_l)\big[A_l(i,j)\Omega_l((i,j),Y)A_l^T(i,j)$$
$$+ \epsilon_l^{-1}(i,j)N_l(i,j)N_l^T(i,j)\big] + \Lambda_l((i,j),Y). \tag{3.31}$$

Under Assumption 3.3, it follows from (3.29)–(3.31) that

$$\tilde{f}_l((i,j),X) \leq (1+\mu_l)\big[A_l(i,j)\Omega_l((i,j),X)A_l^T(i,j)$$
$$+ \epsilon_l^{-1}(i,j)N_l(i,j)N_l(i,j)^T\big] + \Lambda_l((i,j),Y)$$
$$< (1+\mu_l)\big[A_l(i,j)\Omega_l((i,j),Y)A_l^T(i,j)$$
$$+ \epsilon_l^{-1}(i,j)N_l(i,j)N_l^T(i,j)\big] + \Lambda_l((i,j),Y)$$
$$= \tilde{g}_l((i,j),Y).$$

Moreover, the symmetric property of matrix functions $\tilde{f}_l((i,j),X)$ and $\tilde{g}_l((i,j),Y)$ ($l=1,2$) can be observed directly from their expressions. It is now evident to see that the established matrix functions $\tilde{f}_l((i,j),X)$ and $\tilde{g}_l((i,j),Y)$ ($l=1,2$) in (3.30) and (3.31) meet all conditions in Lemma 3.4. On the other hand, based on the matrix functions $\tilde{f}_l((i,j),S(i,j))$ and $\tilde{g}_l((i,j),P(i,j))$ ($l=1,2$) given in (3.30) and (3.31), the expressions of $S(i,j)$ and $P(i,j)$ in (3.21) and (3.22) can be respectively rewritten as

$$S(i,j) = \tilde{f}_1((i,j-1),S(i,j-1)) + \tilde{f}_2((i-1,j),S(i-1,j)),$$
$$P(i,j) = \tilde{g}_1((i,j-1),P(i,j-1)) + \tilde{g}_2((i-1,j),P(i-1,j)).$$

Recalling the inequality condition (3.23) and the initial constraints $S(i,0) < P(i,0)$ and $S(0,j) < P(0,j)$, one gets $S(i,j) < P(i,j)$ from Lemma 3.4. Next, construct matrix $\Theta(i,j)$ as

$$\Theta(i,j) \triangleq \begin{bmatrix} P(i,j) & P(i,j) - S(i,j) \\ P(i,j) - S(i,j) & P(i,j) - S(i,j) \end{bmatrix} \tag{3.32}$$

which implies $\Theta(i,j) > 0$. Moreover, inequality (3.10) can be assured from (3.23) owing to

$$\epsilon_l^{-1}(i,j)I - \bar{M}(i,j)\Theta(i,j)\bar{M}^T(i,j)$$
$$= \epsilon_l^{-1}(i,j)I - M(i,j)P(i,j)M^T(i,j) > 0.$$

Substituting (3.25), (3.26), and (3.32) into the right-side of (3.11) and taking the recursions (3.21)–(3.22) into account, it follows from some tedious yet routine manipulations that $\Theta(i,j)$ given in (3.32) is indeed a solution to (3.11). It is now concluded that (3.12) and (3.26) are true according to Theorem 3.1 and Corollary 3.1.

In the third part of this proof, it remains to show that the proposed filter (3.2) with parameters (3.24) and (3.25) is optimal. Recalling (3.11) and (3.32), we have

$$S(i,j) = [I \quad -I]\Theta(i,j)[I \quad -I]^T$$

$$
\begin{aligned}
&= (1+\mu_1)\big[\zeta_1(i,j-1)\Psi_1(i,j-1)\zeta_1^T(i,j-1)\\
&\quad + \epsilon_1^{-1}(i,j-1)\bar{E}_1(i,j-1)\bar{E}_1^T(i,j-1)\big]\\
&\quad + (1+\mu_2)\big[\zeta_2(i-1,j)\Psi_2(i-1,j)\zeta_2^T(i-1,j)\\
&\quad + \epsilon_2^{-1}(i-1,j)\bar{E}_2(i-1,j)\bar{E}_2^T(i-1,j)\big]\\
&\quad + \bar{\Lambda}_1((i,j-1),P(i,j-1))\hat{K}_1^T(i,j-1)\\
&\quad + \bar{\Lambda}_2((i-1,j),P(i-1,j))\hat{K}_2^T(i-1,j)\\
&\quad + \Lambda_1((i,j-1),P(i,j-1)) + \Lambda_2((i-1,j),P(i-1,j)),
\end{aligned}\tag{3.33}
$$

where

$$
\Psi_l(i,j) \triangleq \big[\Theta^{-1}(i,j) - \epsilon_l(i,j)\bar{M}^T(i,j)\bar{M}(i,j)\big]^{-1},
$$
$$
\bar{E}_l(i,j) \triangleq N_l(i,j) - \hat{K}_l(i,j)E(i,j),
$$
$$
\zeta_l(i,j) \triangleq \big[\; A_l(i,j) - \bar{\beta}(i,j)\hat{K}_l(i,j)C(i,j) \quad \bar{\beta}(i,j)\hat{K}_l(i,j)C(i,j) - \hat{A}_l(i,j) \;\big],
$$
$$
\begin{aligned}
\bar{\Lambda}_l((i,j),P(i,j)) &\triangleq \hat{K}_l(i,j)\big(\bar{\beta}(i,j)C(i,j)P(i,j)C^T(i,j)\\
&\quad + \sigma_\tau^2 D(i,j)P(i,j)D^T(i,j) + R(i,j)\big).
\end{aligned}
$$

Denote $S_l(i,j)$ ($l=1,2,3,4$) *as the partial derivatives of* $\mathrm{tr}\{S(i,j)\}$ *with respect to* $\hat{A}_1(i,j-1)$, $\hat{A}_2(i-1,j)$, $\hat{K}_1(i,j-1)$, *and* $\hat{K}_2(i-1,j)$, *respectively. It is calculated from (3.33) that*

$$
S_1(i,j) = -2(1+\mu_1)\zeta_1(i,j-1)\Psi_1(i,j-1)[0\;\;I]^T,\tag{3.34}
$$
$$
S_2(i,j) = -2(1+\mu_2)\zeta_2(i-1,j)\Psi_2(i-1,j)[0\;\;I]^T,\tag{3.35}
$$
$$
\begin{aligned}
S_3(i,j) = &-2(1+\mu_1)\big(\bar{\beta}(i,j-1)\zeta_1(i,j-1)\bar{\Psi}_1(i,j-1)\\
&\quad + \epsilon^{-1}(i,j-1)\bar{E}_1(i,j-1)E^T(i,j-1)\big)\\
&\quad + 2\bar{\Lambda}_1((i,j-1),P(i,j-1)),
\end{aligned}\tag{3.36}
$$
$$
\begin{aligned}
S_4(i,j) = &-2(1+\mu_2)\big(\bar{\beta}(i-1,j)\zeta_2(i-1,j)\bar{\Psi}_2(i-1,j)\\
&\quad + \epsilon^{-1}(i-1,j)\bar{E}_2(i-1,j)E^T(i-1,j)\big)\\
&\quad + 2\bar{\Lambda}_2((i-1,j),P(i-1,j)),
\end{aligned}\tag{3.37}
$$

where

$$
\bar{\Psi}_l(i,j) \triangleq \Psi_l(i,j)[I\;\;-I]^T C^T(i,j).
$$

Letting the derivatives $S_l(i,j)$ ($l=1,2,3,4$) *in (3.34)–(3.37) be zero, one can confirm from some tedious computations that the filter parameters* $\hat{A}_l(i,j)$ *and* $\hat{K}_l(i,j)$ ($l=1,2$) *are indeed given as (3.24) and (3.25). Therefore, the gain parameters designed as (3.24) and (3.25) are the desired ones that locally minimize the upper bound* $S(i,j)$ *in the trace sense. Based on the above discussions, the proof of this theorem is now complete.*

By means of Theorem 3.2, implementation of the developed 2-D recursive filtering scheme is presented in Algorithm 2.

Algorithm 2 The 2-D finite-horizon robust Kalman filtering

Initialization: Give $\Xi(i,0)$, $\Xi(0,j)$, $P(i,0)$, and $P(0,j)$ satisfying $P(i,0) > \Xi(i,0)$, $P(0,j) > \Xi(0,j)$, and (3.23), where i, $j \in [1\ K]$.

1: **for** $i \in [1\ K]$ and $j = 0$ **do**
2:     Calculate $\hat{K}_l(i,0)$ and $\hat{A}_l(i,0)$ from (3.24)–(3.25).
3: **end for**
4: **for** $j \in [1\ K]$ and $i = 0$ **do**
5:     Calculate $\hat{K}_l(0,j)$ and $\hat{A}_l(0,j)$ from (3.24)–(3.25).
6: **end for**
7: Set $i = 1$ and $j = 1$
8: **while** $i \in [1\ K]$ and $j \in [1\ K]$ **do**
9:     Compute $\hat{x}(i,j)$, $S(i,j)$, and $P(i,j)$ from (3.2), (3.21), and (3.22), respectively.
10:     **if** $i \in [1\ K]$ and $j \in [1\ K-1]$ **then**
11:         Verify the validity of (3.23).
12:         **if** (3.23) is valid **then**
13:             Calculate $\hat{K}_l(i,j)$ and $\hat{A}_l(i,j)$ from (3.24)–(3.25).
14:         **else**
15:             Reset the initial conditions and restart.
16:         **end if**
17:     **end if**
18:     **if** $i \in [1\ K]$ and $j = K$ **then**
19:         Set $i = i + 1$ and $j = 1$.
20:     **else**
21:         Set $j = j + 1$.
22:     **end if**
23: **end while**

Remark 3.4 *Inspired by the methodologies developed in [244] for 1-D systems, the robust Kalman filter design scheme is presented for the 2-D shift-varying uncertain systems. According to Corollary 3.1, the generalized estimation error variance is bounded on the condition that there is a 2-D identity quadratic filter for system (3.1). Subsequently, in Theorem 3.2, existence of such a 2-D identity quadratic filter is verified through the established 2-D comparison-like principle. Furthermore, in terms of two sequences of recursive Riccati-like equations, an upper bound is developed for the generalized estimation error variance at each step by resorting to a constructive method, and the (locally) optimal filter gains are also determined to guarantee the tightest upper bound in the trace sense.*

Remark 3.5 *It can be observed from the expression of $S(i,j)$ as shown in (3.21) that increase of the noise variances $Q(i,j)$ and/or $R(i,j)$ gives rise to a larger upper bound $S(i,j)$. On the other hand, the scalar parameters $\bar{\beta}(i,j)$, $\epsilon_1(i,j)$ and $\epsilon_2(i,j)$ involved in the Riccati-like equations (3.21) and (3.22)* *indeed contribute to the feasibility of equations and thus have an influence on the determination of the locally minimal upper bound $S(i,j)$. Particularly, a higher value of $\bar{\beta}(i,j)$ infers that more measurements are available in the sense of expectation, and thus a lower upper bound $S(i,j)$ would be expected. In addition, smaller values of $\epsilon_1(i,j)$ and $\epsilon_2(i,j)$ make condition (3.23) much easier to be satisfied which may, unfortunately, result in a larger $S(i,j)$ as a cost. As such, it makes practical sense to appropriately tune the scaling parameters such that the upper bound $S(i,j)$ is as tight as possible provided the feasibility of (3.21)–(3.23). This trade-off issue between the parameters $\bar{\beta}(i,j)$, $\epsilon_1(i,j)$, $\epsilon_2(i,j)$ and the tightest upper bound $S(i,j)$ is very interesting, which deserves further investigation.*

Remark 3.6 *Under Assumption 3.3, the sufficient condition presented in Theorem 3.2 assures the existence of a 2-D identity quadratic filter, by which the filter parameters can be determined to minimize (in the trace sense) the obtained upper bound of the generalized estimation error variance. It should be pointed out that Assumption 3.3 might be a bit restrictive for certain 2-D shift-varying system. In case that Assumption 3.3 is not satisfied, we could have one more constraint $P(i,j) > S(i,j)$ in Theorem 3.2 so as to guarantee that the matrix $\Theta(i,j)$ constructed in (3.32) is positive definite. In other words, the tightest upper bound and the suboptimal filter can also be designed if there is a set of feasible solutions satisfying (3.21)–(3.23) and $P(i,j) > S(i,j)$.*

Remark 3.7 *It should be noted that computational burden of the filtering algorithm is often a crucial concern in practice. For the considered system (3.1) defined on a finite horizon $[0\ K]$, the parameter dimensions are given as $x(i,j) \in \mathbb{R}^n$, $y(i,j) \in \mathbb{R}^m$, $w(i,j) \in \mathbb{R}^p$, $v(i,j) \in \mathbb{R}^m$ and $F(i,j) \in \mathbb{R}^{n_1 \times n_2}$. With the aid of matrix computations, it can be concluded from the established results that the proposed algorithm has $O(n^3 + n^2(n_1 + n_2 + m + p) + n(n_1 m + n_2^2 + m^2 + p^2) + n_2^3)$ operations at each iteration. Consequently, the whole computational complexity is $O((n^3 + n^2(n_1 + n_2 + m + p) + n(n_1 m + n_2^2 + m^2 + p^2) + n_2^3)K^2)$, which increases polynomially with the growth of the recursive steps and the dimensions of the system parameters. It is worth mentioning that research on matrix computations is a quite active field in applied mathematics and optimization, where many effective and fast algorithms have been introduced such as the Strassen algorithm and the Coppersmith-Winograd algorithm, hence substantial speedups can be expected in the near future.*

3.4 Numerical Example

In this section, a simulation example is given to illustrate the application of the established filtering scheme. It is known that, in the real world, some dynamical processes in air drying, gas absorption and water stream heating can be described by the following Darboux equation [83]:

$$\frac{\partial^2 u(z,t)}{\partial z \partial t} = a_1 \frac{\partial u(z,t)}{\partial t} + a_2 \frac{\partial u(z,t)}{\partial z} + a_0 u(z,t) + b_0 f(z,t),$$

where $u(z,t)$ is an unknown function at space $z \in [0, Z]$ and time $t \in [0, T]$, $f(z,t)$ is a given input function satisfying the state feedback law $f(z,t) = b_1 u(z,t)$, a_0, a_1, a_2, b_0, and b_1 are known real coefficients. By defining $r(z,t) = \frac{\partial u(z,t)}{\partial t} - a_2 u(z,t)$, one has

$$\frac{\partial r(z,t)}{\partial z} = a_1 \frac{\partial u(z,t)}{\partial t} + (a_0 + b_0 b_1) u(z,t), \tag{3.38a}$$

$$\frac{\partial u(z,t)}{\partial t} = r(z,t) + a_2 u(z,t). \tag{3.38b}$$

To discretize (3.38), the following approximations are utilized

$$\frac{\partial r(z,t)}{\partial z} = \frac{r((i+1)\Delta z, j\Delta t) - r(i\Delta z, j\Delta t)}{\Delta z},$$

$$\frac{\partial u(z,t)}{\partial t} = \frac{u(i\Delta z, (j+1)\Delta t) - u(i\Delta z, j\Delta t)}{\Delta t}$$

where Δz and Δt are the step sizes. Denote

$$r(i,j) \triangleq r(i\Delta z, j\Delta t), \quad u(i,j) \triangleq u(i\Delta z, j\Delta t)$$

$$x(i,j) \triangleq [r^T(i,j) \ u^T(i,j)]^T.$$

The equation (3.38) can be cast into

$$x(i,j) = A_1 x(i, j-1) + A_2 x(i-1, j)$$

with $\bar{a}_0 = (a_0 + a_1 a_2 + b_0 b_1) \Delta z$ and

$$A_1 = \begin{bmatrix} 0 & 0 \\ \Delta t & 1 + a_2 \Delta t \end{bmatrix}, \quad A_2 = \begin{bmatrix} 1 + a_1 \Delta z & \bar{a}_0 \\ 0 & 0 \end{bmatrix}.$$

The 2-D discrete-time system is capable of approximating to the original equation (3.38) by properly choosing the step sizes Δz and Δt.

As an important modeling in industry processes, Darboux equation is unavoidably subject to stochastic noises and parameter variances. Particularly, additive noises might originate from poisoning of the reactor and/or fouling of the drying oven/cooling coil. Multiplicative noises may stem from speckle, dark spots or shadows induced possibly by dust in the reactor or sensor. In addition, due to unpredictable environmental changes, system parameters are likely to be varying thereby reflecting parameter variations influenced by certain heterogeneous media. Parameters may also be affected from uncertainties as lack of accuracy. Additionally, measurement degradations should be considered in the procedure of data gathering when takeing unexpected sensor aging, intermittent failures and randomly changed working conditions into account. In this regard, the target system described by (3.1) is considered with the following parameters:

$$\Delta z = \Delta t = 0.1, \quad a_0 = -20.1, \quad a_1 = -5.5,$$

$$a_2 = -4.2, \quad b_0 = 3, \quad b_1 = -2,$$

$$A_1(i,j) = \begin{bmatrix} 0 & 0 \\ 0.1 & 0.58 + 0.1\sin(i)\cos(j) \end{bmatrix},$$

$$\check{A}_1(i,j) = \begin{bmatrix} 0.1e^{-j} & 0.05\sin(i) \\ 0 & 0.12 \end{bmatrix},$$

$$A_2(i,j) = \begin{bmatrix} 0.45 & -0.3 + 0.1\sin(2j) \\ 0 & 0 \end{bmatrix},$$

$$\check{A}_2(i,j) = \begin{bmatrix} 0.1 + 0.05\cos(i) & -0.1 \\ 0.15e^{-3j} & 0 \end{bmatrix},$$

$$B_1(i,j) = \begin{bmatrix} -0.1 \\ 0.2 + 0.1e^{-i} \end{bmatrix}, \quad M(i,j) = [-0.1 \; 0.1],$$

$$B_2(i,j) = \begin{bmatrix} 0.3 - 0.12e^{-2j} \\ -0.4 \end{bmatrix}, \quad E(i,j) = \begin{bmatrix} 0.05 \\ -0.1 \end{bmatrix},$$

$$N_1(i,j) = \begin{bmatrix} 0.15 \\ -0.16 \end{bmatrix}, \quad N_2(i,j) = \begin{bmatrix} 0.25 \\ -0.1 \end{bmatrix},$$

$$C(i,j) = \begin{bmatrix} 0.3 & 0.12 \\ -0.2 & 0.15 \end{bmatrix}, \quad D(i,j) = \begin{bmatrix} -0.1 & 0.16 \\ 0.18 & 0.1 \end{bmatrix}.$$

The uncorrelated noises $\xi(i,j)$, $\tau(i,j)$, $w(i,j)$, and $v(i,j)$ are supposed to be zero-mean Gaussian white noise sequences with variance $\sigma_\xi^2 = 0.09$, $\sigma_\tau^2 = 0.25$, $Q(i,j) = 0.36$, and $R(i,j) = 0.16I$, respectively. For simulation purpose, the initial conditions are given as $\Xi(i,0) = 0.12I$ and $\Xi(0,j) = 0.1I$ for $i,j \in [1 \; K]$ with $K = 50$. The probability mass function of $\beta(i,j)$ is described as

$$p_s(i,j) = \begin{cases} 0, & s = 0 \\ 0.1, & s = 0.5 \\ 0.9, & s = 1 \end{cases}$$

with expectation $\bar{\beta}(i,j) = 0.95$ and variance $\hat{\beta}(i,j) = 0.0225$. Other parameters are set as $F(i,j) = \sin(0.2i)\cos(j)$, $\mu = 1$, and $\epsilon_1(i,j) = \epsilon_2(i,j) \equiv 10$.

For simplicity, the first elements of the unmeasured state and the state estimate are denoted as $x^{(1)}(i,j)$ and $\hat{x}^{(1)}(i,j)$, respectively; the first diagonal elements of the actual generalized estimation error variance and the obtained upper bound are denoted as $\tilde{\Xi}_{11}(i,j)$ and $S_{11}(i,j)$, respectively.

According to the established algorithm, the filter gains can be recursively calculated from (3.24)–(3.25) as listed in Table 3.1 (only part of the results are presented here for space consideration). The corresponding simulation results are shown in Figs. 3.1–3.5 under the given system parameters. Particularly, Figs. 3.1–3.3 show trajectories of the state $x^{(1)}(i,j)$, its estimate $\hat{x}^{(1)}(i,j)$, and the estimation error $x^{(1)}(i,j) - \hat{x}^{(1)}(i,j)$, respectively. Figure 3.4 plots the first diagonal element $\tilde{\Xi}_{11}(i,j)$ of the generalized estimation error variance $\tilde{\Xi}(i,j)$ from 500 independent runs based on the Monte-Carlo method. Figure 3.5 shows the trajectory of the corresponding upper bound $S_{11}(i,j)$. It is seen from Figs. 3.1–3.5 that the proposed 2-D recursive filter achieves the design objective.

TABLE 3.1: Part of the filter gains $\hat{A}_l(i,j)$ and $\hat{K}_l(i,j)$ $(l = 1,2)$

Instant	$(i,j) = (1,2)$		$(i,j) = (2,1)$		$\cdots$
$\hat{A}_1(i,j)$	-0.0001	0.0001	-0.0001	0.0001	$\cdots$
	0.0876	0.5574	0.0866	0.6426	
$\hat{A}_2(i,j)$	0.4625	-0.2735	0.4638	-0.2939	$\cdots$
	0.0001	-0.0001	0.0001	-0.0001	
$\hat{K}_1(i,j)$	0.0071	-0.0154	0.0071	-0.0165	$\cdots$
	0.0681	0.2106	0.1267	0.1241	
$\hat{K}_2(i,j)$	0.0843	-0.2767	0.1479	-0.3079	$\cdots$
	-0.0053	0.0106	-0.0047	0.0103	

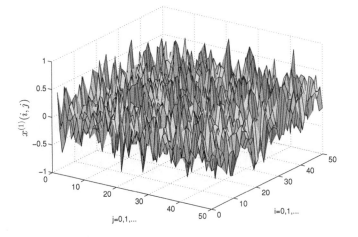

FIGURE 3.1: State evolution of $x^{(1)}(i,j)$ with $\bar{\beta}(i,j) = 0.95$.

In what follows, we shall consider the influence of measurement degradation as well as the scaling parameters $\epsilon_l(i,j)$ $(l = 1,2)$ on the filter

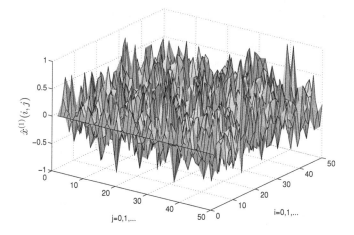

FIGURE 3.2: Estimate evolution of $\hat{x}^{(1)}(i,j)$ with $\bar{\beta}(i,j) = 0.95$.

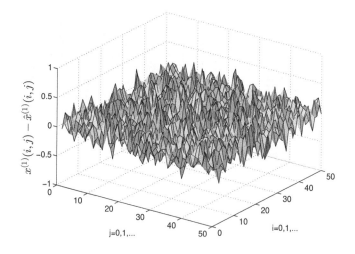

FIGURE 3.3: Estimation error of $x^{(1)}(i,j) - \hat{x}^{(1)}(i,j)$ with $\bar{\beta}(i,j) = 0.95$.

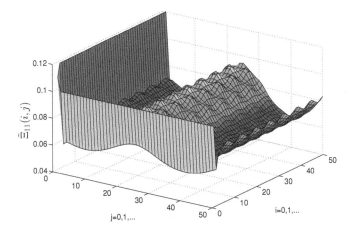

FIGURE 3.4: Evolution of $\tilde{\Xi}_{11}(i,j)$ with $\bar{\beta}(i,j) = 0.95$.

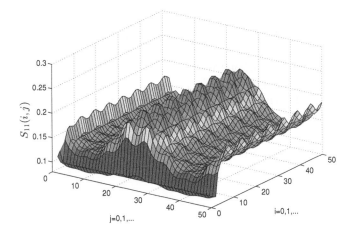

FIGURE 3.5: Evolution of $S_{11}(i,j)$ with $\bar{\beta}(i,j) = 0.95$.

performance. Let us reset the probability mass function of $\beta(i,j)$ as

$$p_s(i,j) = \begin{cases} 0.045, & s = 0 \\ 0.910, & s = 0.5 \\ 0.045, & s = 1 \end{cases}$$

with the other parameters remaining the same as above. It is easy to obtain $\bar{\beta}(i,j) = 0.5$ and $\hat{\beta}(i,j) = 0.0225$. In this case, evolution of $S_{11}(i,j)$ is depicted in Fig. 3.6. On the other hand, we reselect the scaling parameters as $\epsilon_1(i,j) = \epsilon_2(i,j) = 1$ without changing the other parameters given above. The simulation result of this case is depicted in Fig. 3.7.

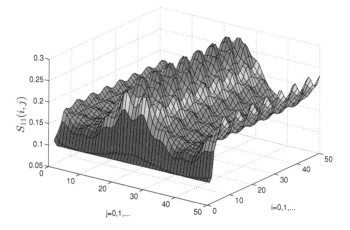

FIGURE 3.6: Evolution of $S_{11}(i,j)$ with $\bar{\beta}(i,j) = 0.5$.

By comparing Fig. 3.5 with Figs. 3.6–3.7, we can conclude that 1) the local minimum upper bound $S_{11}(i,j)$ becomes bigger with the decreasing of the expectation for the measurement degradation; 2) $S_{11}(i,j)$ is influenced with the changing of the parameters $\epsilon_l(i,j)$ $(l = 1, 2)$.

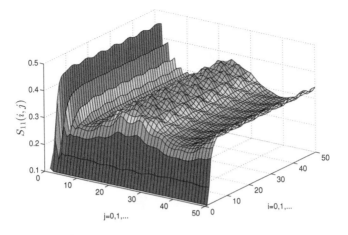

FIGURE 3.7: Evolution of $S_{11}(i,j)$ with $\epsilon_l(i,j) \equiv 1$.

3.5 Summary

This chapter has made the first few attempts to address the finite-horizon robust Kalman filtering problem for 2-D shift-varying systems with norm-bounded parameter uncertainties and measurement degradations. The phenomenon of degraded measurement occurs in a random way, which has been described by a sequence of stochastic variables satisfying the prescribed statistical characterizations. Based on the inductive approach and the established 2-D comparison-like principle, an upper bound for the generalized estimation error variance has been constructed firstly and then the locally minimal upper bound has been obtained with the designed filter whose parameters can be calculated by solving Riccati-like equations suitable for stylized computation. A numerical example of practical insight has also been provided to show the validity of the developed results.

4

Robust Finite-Horizon Filtering for 2-D Systems with Randomly Varying Sensor Delays

The 2-D systems have been receiving increasing popularity for their promising application insights in a variety of practical situations. As for the underlying systems, parameter uncertainties or variations are frequently encountered. An attractive topic of the 2-D uncertain systems has been the robust filtering problem which has drawn growing research attention because of its extensive engineering applications.

For filtering problems with networked communications, the measurements might suffer from unexpected phenomena owing to the inevitable physical limitations such as network bandwidth. The sensor delays have been recognized to be a frequently occurred phenomenon resulting mainly from limited network bandwidth during the signal transmissions. Another issue deserving specific attention is the missing measurements that might contain noises alone. Up to now, the filtering issues with randomly varying sensor delays and/or missing measurements have received considerable interest, and a number of excellent results have been reported in the literature. Nonetheless, a thorough literature search has revealed that the 2-D filtering problem with incomplete measurements has not been fully investigated yet despite the theoretical importance and practical significance.

Concluding the discussions made so far, there is a lack of relevant results on the robust Kalman filtering problem for 2-D shift-varying systems subject to randomly occurring sensor delays and missing measurements. Such a gap seems to result from the following identified technical difficulties: (1) how to analyze the dynamics of 2-D systems in the presence of shift-varying characteristics, parameter uncertainties, and incomplete measurements? (2) how to develop an effective yet novel method suitable for 2-D uncertain systems with sensor delays such that an upper bound could be guaranteed for the estimation error variance? and (3) how to design a 2-D recursive filter with properly designed gain parameters so as to minimize the obtained upper bound? To this end, the primary aim of this chapter is to deal with these essential challenges to facilitate the investigation on 2-D robust filtering scheme.

This chapter addresses the robust finite-horizon filtering problem for 2-D shift-varying systems. The unknown but norm-bounded uncertainties enter into both the system state and the process noise matrices. The considered incomplete measurements embody random sensor delays and missing measurements, which are expressed in a unified way by using a Kronecker delta function. The novelties of this chapter can be highlighted as follows: (1) the proposed 2-D shift-varying system is quite comprehensive which covers parameter uncertainties, randomly varying sensor delays, and missing measurements; (2) a constructive method is provided in the 2-D space to ensure the existence of an upper bound on the estimation error; and (3) a recursive filter is determined that minimizes the obtained upper bound of the estimation error by properly selecting the gain parameters.

The rest of this chapter is arranged as follows. Section 4.1 formulates the 2-D robust filtering problem to be addressed. Section 4.2 provides some useful preliminary knowledge. Section 4.3 presents the existence of an upper bound on the error variance of the state estimation and also proposes the derivable filtering algorithm. A numerical example is employed in Section 4.4 to show the effectiveness of the developed results. Conclusions are drawn finally in Section 4.5.

4.1   Problem Formulation

Consider the following 2-D shift-varying uncertain system defined on a finite horizon $i, j \in [0\ K]$:

$$\vec{x}(i+1, j+1) = (\vec{A}_1(i+1, j) + \Delta \vec{A}_1(i+1, j))\vec{x}(i+1, j)$$
$$+ (\vec{A}_2(i, j+1) + \Delta \vec{A}_2(i, j+1))\vec{x}(i, j+1)$$
$$+ (\vec{B}_1(i+1, j) + \Delta \vec{B}_1(i+1, j))w(i+1, j)$$
$$+ (\vec{B}_2(i, j+1) + \Delta \vec{B}_2(i, j+1))w(i, j+1), \qquad (4.1a)$$
$$\vec{y}(i, j) = \vec{C}(i, j)\vec{x}(i, j), \qquad (4.1b)$$

where $\vec{x}(i, j) \in \mathbb{R}^n$ is the system state, $\vec{y}(i, j) \in \mathbb{R}^m$ is the ideal measurement, and $w(i, j) \in \mathbb{R}^p$ is the process white noise. $\vec{A}_l(i, j)$, $\vec{B}_l(i, j)$ $(l = 1, 2)$, and $\vec{C}(i, j)$ are known shift-varying matrices with compatible dimensions, while $\Delta \vec{A}_l(i, j)$ and $\Delta \vec{B}_l(i, j)$ $(l = 1, 2)$ are uncertain matrices with the following structure:

$$\Delta \vec{A}_l(i, j) = \vec{N}_l(i, j)F(i, j)\vec{M}(i, j),$$
$$\Delta \vec{B}_l(i, j) = \vec{H}_l(i, j)F(i, j)E(i, j),$$

where $\vec{N}_l(i, j)$, $\vec{H}_l(i, j)$, $\vec{M}(i, j)$, and $E(i, j)$ are known shift-varying matrices, and $F(i, j)$ indicates the parameter uncertainty satisfying $F(i, j)F^T(i, j) \leq I$.

Note that the system measurements are often subject to packet dropouts and sensor delays, and the ideal measurements are therefore unavailable. In this case, the actually received system measurements are of the following form:

$$y(i,j) = \delta(\beta(i,j),1)\vec{y}(i,j) + \delta(\beta(i,j),2)\vec{y}(i,j-1)$$
$$+ \delta(\beta(i,j),3)\vec{y}(i-1,j) + v(i,j), \qquad (4.2)$$

where $y(i,j) \in \mathbb{R}^m$ is the received measurement output, $v(i,j) \in \mathbb{R}^m$ is the measurement white noise, and $\delta(\beta(i,j),k)$ with $k \in [1\ 4]$ is the Kronecker delta function taking values 1 for $\beta(i,j) = k$ and 0 otherwise. The stochastic variable $\beta(i,j)$ obeys the following probability distribution:

$$\mathrm{Prob}\{\beta(i,j) = k\} = \bar{\beta}_k(i,j)$$

where $\bar{\beta}_k(i,j)$ belonging to $[0,1]$ are given constants with $\sum_{k=1}^{4} \bar{\beta}_k(i,j) = 1$.

Remark 4.1 *By introducing the Kronecker delta function, the actual measurement described by (4.2) is capable of characterizing different types of incomplete measurements when choosing different values for the stochastic variable $\beta(i,j)$. To be more specific, if $\beta(i,j) = 1$, then the received observation is exactly presented as $y(i,j) = \vec{y}(i,j) + v(i,j)$, which infers that the sensor works normally; if $\beta(i,j) = 2$ or 3, the measurement suffers from the sensor delay; and if $\beta(i,j) = 4$, the sensor receives only the measurement noise. In this way, a unified form is established for the incomplete measurements which cover not only the randomly varying sensor delays but also the missing measurements, where the corresponding occurrences depend on the given probability distribution of $\beta(i,j)$. Such a kind of incomplete measurement will definitely increase the mathematical complexities and difficulties in analyzing the system dynamics.*

The initial conditions $\{\vec{x}(i,0)\}_{i=1}^{K}$ and $\{\vec{x}(0,j)\}_{j=1}^{K}$ for system (4.1) are two uncorrelated white sequences satisfying $\mathbb{E}\{\vec{x}(i,0)\} = \vec{u}_1(i)$ and $\mathbb{E}\{\vec{x}(0,j)\} = \vec{u}_2(j)$, where $\vec{u}_1(i)$ and $\vec{u}_2(j)$ are known vectors for all $i,j \in [1\ K]$. For simplicity, we set $\vec{x}(i,j) \equiv 0$ if either i or j is negative and $\vec{x}(0,0) = 0$ throughout this chapter. Moreover, we make the following assumptions.

Assumption 4.1 *For $i,j \in [0\ K]$, the random variable $\beta(i,j)$ is unrelated with the process and measurement noises $w(i,j)$ and $v(i,j)$ as well as the initial conditions $\vec{x}(i,0)$ and $\vec{x}(0,j)$.*

Assumption 4.2 *For $i,j,k,l \in [0\ K]$, one has the following statistical relationship:*

$$\mathbb{E}\left\{\eta(i,j)\eta^T(k,l)\right\} = \mathrm{diag}\left\{\vec{\Xi}(i,0),\vec{\Xi}(0,j),Q(i,j),R(i,j)\right\}\delta(i,k)\delta(j,l),$$

where $\eta(i,j) \triangleq [\vec{x}^T(i,0)\ \vec{x}^T(0,j)\ w^T(i,j)\ v^T(i,j)]^T$, matrices $\vec{\Xi}(i,0), \vec{\Xi}(0,j) \geq 0$ and $Q(i,j), R(i,j) > 0$ are the known second-order moments of initial states and noises, respectively.

Define

$$x(i,j) \triangleq [\vec{x}^T(i,j) \; \vec{x}^T(i,j-1) \; \vec{x}^T(i-1,j)]^T,$$

$$A_1(i,j) \triangleq \begin{bmatrix} \vec{A}_1(i,j) & 0 & 0 \\ I & 0 & 0 \\ 0 & 0 & 0 \end{bmatrix}, \quad A_2(i,j) \triangleq \begin{bmatrix} \vec{A}_2(i,j) & 0 & 0 \\ 0 & 0 & 0 \\ I & 0 & 0 \end{bmatrix},$$

$$B_l(i,j) \triangleq [\vec{B}_l^T(i,j) \; 0 \; 0]^T, \quad N_l(i,j) \triangleq [\vec{N}_l^T(i,j) \; 0 \; 0]^T,$$

$$H_l(i,j) \triangleq [\vec{H}_l^T(i,j) \; 0 \; 0]^T, \quad M(i,j) \triangleq [\vec{M}(i,j) \; 0 \; 0],$$

$$\Delta A_l(i,j) \triangleq N_l(i,j)F(i,j)M(i,j),$$

$$\Delta B_l(i,j) \triangleq H_l(i,j)F(i,j)E(i,j), \quad l = 1,2$$

$$\Lambda_\beta(i,j) \triangleq \text{diag}_{1\le k\le 3}\{\delta(\beta(i,j),k)I\},$$

$$C(i,j) \triangleq [\vec{C}(i,j) \; \vec{C}(i,j-1) \; \vec{C}(i-1,j)].$$

Then, the following equivalent compact form for (4.1) and (4.2) is derived:

$$\begin{aligned} x(i+1,j+1) &= [A_1(i+1,j) + \Delta A_1(i+1,j)]x(i+1,j) \\ &\quad + [A_2(i,j+1) + \Delta A_2(i,j+1)]x(i,j+1) \\ &\quad + [B_1(i+1,j) + \Delta B_1(i+1,j)]w(i+1,j) \\ &\quad + [B_2(i,j+1) + \Delta B_2(i,j+1)]w(i,j+1), \quad\quad (4.3a) \\ y(i,j) &= C(i,j)\Lambda_\beta(i,j)x(i,j) + v(i,j). \quad\quad\quad\quad\quad (4.3b) \end{aligned}$$

In this chapter, a recursive filter for (4.3) is proposed as follows:

$$\begin{aligned} &\hat{x}(i+1,j+1) \\ &= \hat{A}_1(i+1,j)\hat{x}(i+1,j) + \hat{A}_2(i,j+1)\hat{x}(i,j+1) \\ &\quad + \hat{K}_1(i+1,j)\left(y(i+1,j) - C(i+1,j)\bar{\Lambda}_\beta(i+1,j)\hat{x}(i+1,j)\right) \\ &\quad + \hat{K}_2(i,j+1)\left(y(i,j+1) - C(i,j+1)\bar{\Lambda}_\beta(i,j+1)\hat{x}(i,j+1)\right), \quad (4.4) \end{aligned}$$

where $\hat{x}(i,j) \in \mathbb{R}^{3n}$ is the state estimation, $\bar{\Lambda}_\beta(i,j) \triangleq \text{diag}_{1\le k\le 3}\{\bar{\beta}_k(i,j)I\}$, $\hat{A}_l(i,j)$ and $\hat{K}_l(i,j)$ ($l = 1,2$) are shift-varying filter parameters to be designed, and the initial condition associated with (4.4) is given as $\hat{x}(i,0) = \hat{x}(0,j) \equiv 0$ for $i,j \in [0 \; K]$.

The purpose of this chapter is to design a filter of form (4.4) such that, in the presence of all parameter uncertainties, randomly varying sensor delays, and missing measurement, the following two requirements are simultaneously satisfied:

R1) there is a sequence of positive definite matrices $\{S(i,j)\}_{i,j=0}^K$ satisfying

$$\mathbb{E}\left\{[x(i,j) - \hat{x}(i,j)][x(i,j) - \hat{x}(i,j)]^T\right\} \le S(i,j), \quad\quad (4.5)$$

namely, an upper bound $S(i,j)$ can be derived for the second-order moment of the estimation error (or the general estimation error variance);

R2) the obtained upper bound is locally minimized by appropriately designing the filter gain parameters.

The problem under consideration is referred to as the 2-D robust filtering problem over a finite horizon. Some preliminaries are given in the next section to further deal with the proposed problem.

4.2 Preliminaries

In this section, some basic matrix inequalities are presented. In addition, to facilitate the design of the 2-D robust filter (4.4), the augmented system and its corresponding quadratic filter are given as well.

To begin with, the following lemmas are introduced for preparation.

Lemma 4.1 *Denote a random matrix as*

$$\tilde{\Lambda}_\beta(i,j) \triangleq \text{diag}_{1 \le k \le 3} \{\tilde{\beta}_k(i,j)I\} \tag{4.6}$$

with $\tilde{\beta}_k(i,j) = \delta(\beta(i,j),k) - \bar{\beta}_k(i,j)$ *and let* X *be a real-valued matrix with appropriate dimensions. Then, one has*

$$\mathbb{E}\{\tilde{\Lambda}_\beta(i,j)X\tilde{\Lambda}_\beta^T(i,j)\} = \hat{\Lambda}_\beta(i,j) \circ X, \tag{4.7}$$

where

$$\hat{\Lambda}_\beta(i,j) = \begin{bmatrix} \hat{\beta}_{11}(i,j) & \hat{\beta}_{12}(i,j) & \hat{\beta}_{13}(i,j) \\ \hat{\beta}_{21}(i,j) & \hat{\beta}_{22}(i,j) & \hat{\beta}_{23}(i,j) \\ \hat{\beta}_{31}(i,j) & \hat{\beta}_{32}(i,j) & \hat{\beta}_{33}(i,j) \end{bmatrix} \otimes (\mathbf{1}\mathbf{1}^T)$$

in which

$$\hat{\beta}_{kl}(i,j) = \begin{cases} \bar{\beta}_k(i,j) - \bar{\beta}_k^2(i,j), & k = l \\ -\bar{\beta}_k(i,j)\bar{\beta}_l(i,j), & k \ne l. \end{cases}$$

Proof *According to the definitions of* $\delta(\beta(i,j),k)$ *and* $\bar{\beta}_k(i,j)$*, it is easy to obtain that*

$$\mathbb{E}\{\tilde{\beta}_k^2(i,j)\} = \bar{\beta}_k(i,j) - \bar{\beta}_k^2(i,j),$$
$$\mathbb{E}\{\tilde{\beta}_k(i,j)\tilde{\beta}_l(i,j)\} = -\bar{\beta}_k(i,j)\bar{\beta}_l(i,j), \quad k \ne l.$$

Recalling the expression of $\hat{\Lambda}_\beta(i,j)$*, the validity of (4.7) follows directly.*

Lemma 4.2 *For matrices* $A \ge 0$ *and* $B \ge 0$ *with appropriate dimensions, one has*

$$A \circ B \ge 0.$$

Lemma 4.3 *For matrices $A \in \mathbb{R}^{n \times n}$ and $B \in \mathbb{R}^{m \times m}$, assume that $\lambda_1, \lambda_2, \ldots, \lambda_n$ are the eigenvalues of A and $\sigma_1, \sigma_2, \ldots, \sigma_m$ are those of B. Then, the eigenvalues of $A \otimes B$ are given as $\lambda_i \sigma_j$ for $i \in [1\ n]$ and $j \in [1\ m]$.*

In what follows, the evolution of the second-order moment of $x(i,j)$ is to be derived firstly, and then an augmented system will be presented by introducing a new vector. Subsequently, a novel filter definition is to be proposed to address the 2-D robust filtering problem.

By defining $\Xi(i,j) \triangleq \mathbb{E}\{x(i,j)x^T(i,j)\}$ and recalling (4.3), one has the evolution of $\Xi(i,j)$ as

$$
\begin{aligned}
\Xi(i+1,j+1) = \; & f_1((i+1,j), \Xi(i+1,j), Q(i+1,j)) \\
& + f_2((i,j+1), \Xi(i,j+1), Q(i,j+1)) \\
& + f_3(x(i+1,j), x(i,j+1)),
\end{aligned}
\tag{4.8}
$$

where, for a given matrix X and $l = 1, 2$,

$$
\begin{aligned}
& f_l((i,j), X, Q(i,j)) \\
& \triangleq [A_l(i,j) + \Delta A_l(i,j)]\, X\, [A_l(i,j) + \Delta A_l(i,j)]^T \\
& \quad + [B_l(i,j) + \Delta B_l(i,j)]\, Q(i,j)\, [B_l(i,j) + \Delta B_l(i,j)]^T, \\
& f_3(x(i+1,j), x(i,j+1)) \\
& \triangleq \Gamma(x(i+1,j), x(i,j+1)) + \Gamma^T(x(i+1,j), x(i,j+1)), \\
& \Gamma(x(i+1,j), x(i,j+1)) \\
& \triangleq \mathbb{E}\big\{[A_1(i+1,j) + \Delta A_1(i+1,j)]x(i+1,j)x^T(i,j+1) \\
& \quad \times [A_2(i,j+1) + \Delta A_2(i,j+1)]^T\big\}.
\end{aligned}
$$

Define $\bar{x}(i,j) \triangleq [x^T(i,j)\ \hat{x}^T(i,j)]^T$. It follows from (4.3) and (4.4) that

$$
\begin{aligned}
\bar{x}(i+1,j+1) = \; & \big[\bar{A}_1(i+1,j) + \Delta\bar{A}_1(i+1,j)\big]\, \bar{x}(i+1,j) \\
& + \big[\bar{A}_2(i,j+1) + \Delta\bar{A}_2(i,j+1)\big]\, \bar{x}(i,j+1) \\
& + \big[\bar{B}_1(i+1,j) + \Delta\bar{B}_1(i+1,j)\big]\, \bar{w}(i+1,j) \\
& + \big[\bar{B}_2(i,j+1) + \Delta\bar{B}_2(i,j+1)\big]\, \bar{w}(i,j+1),
\end{aligned}
\tag{4.9}
$$

where

$$
\begin{aligned}
& \bar{A}_l(i,j) = \begin{bmatrix} A_l(i,j) & 0 \\ \hat{K}_l(i,j)C(i,j)\bar{\Lambda}_\beta(i,j) & \check{A}_l(i,j) \end{bmatrix}, \\
& \check{A}_l(i,j) = \hat{A}_l(i,j) - \hat{K}_l(i,j)C(i,j)\bar{\Lambda}_\beta(i,j), \\
& \bar{N}_l(i,j) = \begin{bmatrix} N_l(i,j) \\ 0 \end{bmatrix}, \quad \bar{H}_l(i,j) = \begin{bmatrix} H_l(i,j) \\ 0 \end{bmatrix}, \\
& \bar{M}(i,j) = [M(i,j)\ 0], \quad \bar{E}(i,j) = [E(i,j)\ 0],
\end{aligned}
$$

$$\bar{B}_l(i,j) = \begin{bmatrix} B_l(i,j) & 0 \\ 0 & \hat{K}_l(i,j) \end{bmatrix},$$

$$\Delta\bar{A}_l(i,j) = \bar{N}_l(i,j)F(i,j)\bar{M}(i,j),$$

$$\Delta\bar{B}_l(i,j) = \bar{H}_l(i,j)F(i,j)\bar{E}(i,j), \quad l = 1,2$$

$$\bar{w}(i,j) = \begin{bmatrix} w(i,j) \\ C(i,j)\tilde{\Lambda}_\beta(i,j)x(i,j) + v(i,j) \end{bmatrix}.$$

By further denoting $\bar{\Theta}(i,j) \triangleq \mathbb{E}\{\bar{x}(i,j)\bar{x}^T(i,j)\}$ and employing the uncorrelatedness of stochastic variables $w(i,j)$, $v(i,j)$, and $\beta(i,j)$, the evolution of $\bar{\Theta}(i,j)$ is governed by the following equality:

$$\bar{\Theta}(i+1,j+1)$$

$$= \left[\bar{A}_1(i+1,j) + \Delta\bar{A}_1(i+1,j)\right]\bar{\Theta}(i+1,j)\left[\bar{A}_1(i+1,j) + \Delta\bar{A}_1(i+1,j)\right]^T$$

$$+ \left[\bar{A}_2(i,j+1) + \Delta\bar{A}_2(i,j+1)\right]\bar{\Theta}(i,j+1)\left[\bar{A}_2(i,j+1) + \Delta\bar{A}_2(i,j+1)\right]^T$$

$$+ \left[\bar{B}_1(i+1,j) + \Delta\bar{B}_1(i+1,j)\right]\bar{Q}(i+1,j)\left[\bar{B}_1(i+1,j) + \Delta\bar{B}_1(i+1,j)\right]^T$$

$$+ \left[\bar{B}_2(i,j+1) + \Delta\bar{B}_2(i,j+1)\right]\bar{Q}(i,j+1)\left[\bar{B}_2(i,j+1) + \Delta\bar{B}_2(i,j+1)\right]^T$$

$$+ \Phi(\bar{x}(i+1,j),\bar{x}(i,j+1)) + \Phi^T(\bar{x}(i+1,j),\bar{x}(i,j+1)), \tag{4.10}$$

where Lemma 4.1 has been utilized and

$$\Phi(\bar{x}(i+1,j),\bar{x}(i,j+1)) \triangleq \mathbb{E}\left\{[\bar{A}_1(i+1,j) + \Delta\bar{A}_1(i+1,j)]\bar{x}(i+1,j)\right.$$

$$\left. \times \bar{x}^T(i,j+1)[\bar{A}_2(i,j+1) + \Delta\bar{A}_2(i,j+1)]^T\right\},$$

$$\bar{Q}(i,j) \triangleq \begin{bmatrix} Q(i,j) & 0 \\ 0 & \bar{R}(i,j) \end{bmatrix}$$

in which

$$\bar{R}(i,j) = C(i,j)\left(\hat{\Lambda}_\beta(i,j) \circ \Xi(i,j)\right)C^T(i,j) + R(i,j).$$

For convenience, we rewrite $\bar{\Theta}(i,j)$ in (4.10) as

$$\bar{\Theta}(i+1,j+1) = \bar{f}_1((i+1,j),\bar{\Theta}(i+1,j),\bar{Q}(i+1,j))$$

$$+ \bar{f}_2((i,j+1),\bar{\Theta}(i,j+1),\bar{Q}(i,j+1))$$

$$+ \bar{f}_3(\bar{x}(i+1,j),\bar{x}(i,j+1)) \tag{4.11}$$

where, for a given matrix X and $l = 1,2$,

$$\bar{f}_l((i,j),X,\bar{Q}(i,j))$$

$$\triangleq \left[\bar{A}_l(i,j) + \Delta\bar{A}_l(i,j)\right]X\left[\bar{A}_l(i,j) + \Delta\bar{A}_l(i,j)\right]^T$$

$$+ \left[\bar{B}_l(i,j) + \Delta\bar{B}_l(i,j)\right]\bar{Q}(i,j)\left[\bar{B}_l(i,j) + \Delta\bar{B}_l(i,j)\right]^T,$$

$$\bar{f}_3(\bar{x}(i+1,j),\bar{x}(i,j+1))$$

$$\triangleq \Phi(\bar{x}(i+1,j),\bar{x}(i,j+1)) + \Phi^T(\bar{x}(i+1,j),\bar{x}(i,j+1)).$$

Remark 4.2 *Since the evolution of $\bar{\Theta}(i,j)$ includes uncertainties $\Delta\bar{A}_l(i,j)$ and $\Delta\bar{B}_l(i,j)$ ($l = 1, 2$), it is impossible to obtain the analytically expression of $\bar{\Theta}(i,j)$. What's worse, this fact further leads to the inaccessibility of a globally optimal filter. Fortunately, we have an alternative method for constructing an upper bound for $\bar{\Theta}(i,j)$ which is then minimized to determine the corresponding filter parameters.*

Inspired by the definition of 1-D quadratic filter introduced in [237], the following 2-D quadratic filter is proposed as an extension for the 2-D shift-varying systems.

Definition 4.1 *The filter (4.4) is said to be a 2-D quadratic filter concerning matrices $\Theta(i,j)$ and $P(i,j)$ for system (4.9) if there are two sequences of positive definite matrices $\{P(i,j)\}_{i,j=0}^K$ and $\{\Theta(i,j)\}_{i,j=0}^K$, and four sequences of positive scalars $\{\mu_l(i,j)\}_{i,j=0}^K$, $\{\epsilon_l(i,j)\}_{i,j=0}^K$ ($l = 1, 2$) such that*

$$\mu_l^{-1}(i,j)I - \bar{E}(i,j)\tilde{Q}(i,j)\bar{E}^T(i,j) > 0, \tag{4.12}$$

$$\epsilon_l^{-1}(i,j)I - \bar{M}(i,j)\Theta(i,j)\bar{M}^T(i,j) > 0, \tag{4.13}$$

$$\begin{aligned}
\Theta(i+1,j+1) = {} & \bar{g}_1((i+1,j),\Theta(i+1,j),\tilde{Q}(i+1,j)) \\
& + \bar{g}_2((i,j+1),\Theta(i,j+1),\tilde{Q}(i,j+1)) \\
& + \bar{g}_3(\Theta(i+1,j),\Theta(i,j+1))
\end{aligned} \tag{4.14}$$

are satisfied, where

$$\tilde{Q}(i,j) \triangleq \mathrm{diag}\{Q(i,j),\tilde{R}(i,j)\},$$

$$\tilde{R}(i,j) \triangleq C(i,j)(\hat{\Lambda}_\beta(i,j) \circ P(i,j))C^T(i,j) + R(i,j)$$

and for a given matrix X, $i,j \in [0\ K]$ and $l = 1, 2$,

$$\bar{g}_l((i,j),X,\tilde{Q}(i,j)) \triangleq g_l((i,j),X) + h_l((i,j),\tilde{Q}(i,j)),$$

$$\begin{aligned}
g_l((i,j),X) \triangleq {} & \bar{A}_l(i,j)\left[X^{-1} - \epsilon_l(i,j)\bar{M}^T(i,j)\bar{M}(i,j)\right]^{-1}\bar{A}_l^T(i,j) \\
& + \epsilon_l^{-1}(i,j)\bar{N}_l(i,j)\bar{N}_l^T(i,j),
\end{aligned}$$

$$\begin{aligned}
h_l((i,j),\tilde{Q}(i,j)) \triangleq {} & \bar{B}_l(i,j)\left[\tilde{Q}^{-1}(i,j) - \mu_l(i,j)\bar{E}^T(i,j)\bar{E}(i,j)\right]^{-1} \\
& \times \bar{B}_l^T(i,j) + \mu_l^{-1}(i,j)\bar{H}_l(i,j)\bar{H}_l^T(i,j),
\end{aligned}$$

$$\begin{aligned}
\bar{g}_3(\Theta(i+1,j),\Theta(i,j+1)) \triangleq {} & \tau g_1((i+1,j),\Theta(i+1,j)) \\
& + \tau^{-1}g_2((i,j+1),\Theta(i,j+1))
\end{aligned}$$

in which $\tau > 0$ is any given constant scalar.

Remark 4.3 *The concept of 2-D quadratic filter proposed in Definition 4.1 will play a significant role in developing an upper bound for the estimation error variance. It will be demonstrated later that, if there is a 2-D quadratic filter in the form of (4.4), then the positive definite matrix $\Theta(i,j)$ satisfying (4.12)–(4.14) is just the expected upper bound of $\bar{\Theta}(i,j)$.*

4.3   Finite-Horizon Robust Kalman Filter Design

In this section, an upper bound for matrix $\bar{\Theta}(i,j)$ is first derived under the condition that a 2-D quadratic filter exists. Subsequently, a sufficient criterion is established for designing the 2-D quadratic filter through a constructive method. Finally, the desired optimal 2-D robust Kalman filter is determined in order to minimize the obtained bound.

The following theorem illustrates the existence of an upper bound of matrix $\bar{\Theta}(i,j)$.

Theorem 4.1 *Assume that (4.4) is a 2-D quadratic filter concerning matrices $\Theta(i,j)$ and $P(i,j)$ for system (4.9). If*

$$\Xi(i,j) \leq P(i,j) \tag{4.15}$$

and the initial constraints

$$\bar{\Theta}(i,0) \leq \Theta(i,0), \quad \bar{\Theta}(0,j) \leq \Theta(0,j)$$

are satisfied, then the following inequality holds:

$$\bar{\Theta}(i,j) \leq \Theta(i,j), \quad i,j \in [0 \ \ K]. \tag{4.16}$$

Proof *For simplicity, set*

$$\Pi_\beta(i,j) = \begin{bmatrix} \hat{\beta}_{11}(i,j) & \hat{\beta}_{12}(i,j) & \hat{\beta}_{13}(i,j) \\ \hat{\beta}_{21}(i,j) & \hat{\beta}_{22}(i,j) & \hat{\beta}_{23}(i,j) \\ \hat{\beta}_{31}(i,j) & \hat{\beta}_{32}(i,j) & \hat{\beta}_{33}(i,j) \end{bmatrix}.$$

According to the definition of $\hat{\beta}_{kl}(i,j)$ $(k,l \in [1 \ \ 3])$ given in Lemma 4.1, it is easy to confirm that all the principle minors of $\Pi_\beta(i,j)$ are nonnegative and then, based on the linear algebra theory, one has

$$\Pi_\beta(i,j) \geq 0. \tag{4.17}$$

In addition, note that $\mathbf{11}^T \geq 0$ and $P(i,j) - \Xi(i,j) \geq 0$, it follows from Lemmas 4.2–4.3 that

$$(P(i,j) - \Xi(i,j)) \circ \hat{\Lambda}_\beta(i,j) \geq 0 \tag{4.18}$$

that is,

$$\hat{\Lambda}_\beta(i,j) \circ \Xi(i,j) \leq \hat{\Lambda}_\beta(i,j) \circ P(i,j) \tag{4.19}$$

which further infers the relationship $\bar{Q}(i,j) \leq \tilde{Q}(i,j)$. Moreover, if there is a *matrix Y satisfying $Y \leq \Theta(i,j)$, then it follows from conditions (4.12), (4.13), and Lemma 3.2 that*

$$\bar{f}_l((i,j), Y, \bar{Q}(i,j)) \leq \bar{f}_l((i,j), \Theta(i,j), \tilde{Q}(i,j))$$
$$\leq \bar{g}_l((i,j), \Theta(i,j), \tilde{Q}(i,j)) \qquad (4.20)$$

for $l = 1, 2$.

In the following, the assertion of this theorem will be confirmed by using the induction method, where the relationship between $\bar{\Theta}(i,j)$ and $\Theta(i,j)$ will be discussed inductively. By applying (4.10), (4.14), (4.20) as well as Lemma 3.3, one has

$$\begin{aligned}
\bar{\Theta}(1,1) &= \bar{f}_1((1,0), \bar{\Theta}(1,0), \bar{Q}(1,0)) + \bar{f}_2((0,1), \bar{\Theta}(0,1), \bar{Q}(0,1)) \\
&\quad + \bar{f}_3(\bar{x}(1,0), \bar{x}(0,1)) \\
&\leq \bar{f}_1((1,0), \Theta(1,0), \tilde{Q}(1,0)) + \bar{f}_2((0,1), \Theta(0,1), \tilde{Q}(0,1)) \\
&\quad + \bar{g}_3(\Theta(1,0), \Theta(0,1)) \\
&\leq \bar{g}_1((1,0), \Theta(1,0), \tilde{Q}(1,0)) + \bar{g}_2((0,1), \Theta(0,1), \tilde{Q}(0,1)) \\
&\quad + \bar{g}_3(\Theta(1,0), \Theta(0,1)) \\
&= \Theta(1,1).
\end{aligned}$$

Hence, $\bar{\Theta}(i,j) \leq \Theta(i,j)$ is valid for $(i,j) \in \{(i_0, j_0)| i_0, j_0 \leq 1; i_0 + j_0 = 2\}$. We now assume that $\bar{\Theta}(i,j) \leq \Theta(i,j)$ holds for all $(i,j) \in \{(i_0, j_0)| i_0, j_0 \in [1 \quad k-1]; i_0 + j_0 = k\}$, where $k \in [2 \quad 2K-1]$ is a given integer. Then, we arrive at

$$\begin{aligned}
\bar{\Theta}(i,j) &= \bar{f}_1((i,j-1), \bar{\Theta}(i,j-1), \bar{Q}(i,j-1)) \\
&\quad + \bar{f}_2((i-1,j), \bar{\Theta}(i-1,j), \bar{Q}(i-1,j)) \\
&\quad + \bar{f}_3(\bar{x}(i,j-1), \bar{x}(i-1,j)) \\
&\leq \bar{f}_1((i,j-1), \Theta(i,j-1), \tilde{Q}(i,j-1)) \\
&\quad + \bar{f}_2((i-1,j), \Theta(i-1,j), \tilde{Q}(i-1,j)) \\
&\quad + \bar{g}_3(\Theta(i,j-1), \Theta(i-1,j)) \\
&\leq \bar{g}_1((i,j-1), \Theta(i,j-1), \tilde{Q}(i,j-1)) \\
&\quad + \bar{g}_2((i-1,j), \Theta(i-1,j), \tilde{Q}(i-1,j)) \\
&\quad + \bar{g}_3(\Theta(i,j-1), \Theta(i-1,j)) \\
&= \Theta(i,j).
\end{aligned}$$

Based on the induction method, the validity of (4.16) is confirmed for all $i, j \in [0 \quad K]$, which ends the proof.

The following result is easily accessible from Theorem 4.1.

Corollary 4.1 *Under the same conditions as in Theorem 4.1, the inequality given below is satisfied:*

$$\mathbb{E}\left\{[x(i,j) - \hat{x}(i,j)][x(i,j) - \hat{x}(i,j)]^T\right\}$$
$$\leq [I - I]\Theta(i,j)[I - I]^T, \quad i,j \in [0 \ K]. \tag{4.21}$$

Remark 4.4 *Theorem 4.1 shows that an upper bound of matrix $\bar{\Theta}(i,j)$ can be determined if there is a 2-D quadratic filter (4.4) satisfying (4.12)–(4.14) provided that $\Xi(i,j) \leq P(i,j)$, and the estimation error variance is then bounded by a certain matrix presented in Corollary 4.1. However, such a bound may not be unique owing to the possible non-unique solution $\Theta(i,j)$ to (4.12)–(4.14). With purpose of reducing the conservatism as much as possible, it is natural to ascertain the minimal upper bound by optimizing the filter gain parameters.*

For $l = 1, 2$, any given matrices X, Y and $Z(i,j)$ satisfying $Y \geq X > 0$ and $Z(i,j) > 0$ as well as a positive scalar τ, we denote the following notations for presentation convenience:

$$\tau_1 \triangleq \tau, \quad \tau_2 \triangleq \tau^{-1},$$

$$\Omega_l((i,j), X) \triangleq \left[X^{-1} - \epsilon_l(i,j)M^T(i,j)M(i,j)\right]^{-1},$$

$$\bar{\Omega}_l((i,j), X) \triangleq \left[\epsilon_l^{-1}(i,j)I - M(i,j)XM^T(i,j)\right]^{-1},$$

$$\Upsilon_l((i,j), X, Y) \triangleq (1 + \tau_l)C(i,j)\bar{\Lambda}_\beta(i,j)\Omega_l((i,j), X)\bar{\Lambda}_\beta(i,j)C^T(i,j)$$
$$+ C(i,j)(\hat{\Lambda}_\beta(i,j) \circ Y)C^T(i,j) + R(i,j),$$

$$\tilde{\Omega}_l((i,j), X, Y) \triangleq (1 + \tau_l)^2 A_l(i,j)\Omega_l((i,j), X)\bar{\Lambda}_\beta(i,j)C^T(i,j)$$
$$\times \Upsilon_l^{-1}((i,j), X, Y)C(i,j)\bar{\Lambda}_\beta(i,j)\Omega_l((i,j), X)A_l^T(i,j),$$

$$d_l((i,j), X) \triangleq A_l(i,j)\Omega_l((i,j), X)A_l^T(i,j) + \epsilon_l^{-1}(i,j)N_l(i,j)N_l^T(i,j),$$

$$e_l((i,j), Q(i,j)) \triangleq B_l(i,j)\left[Q^{-1}(i,j) - \mu_l(i,j)E^T(i,j)E(i,j)\right]^{-1} B_l^T(i,j)$$
$$+ \mu_l^{-1}(i,j)H_l(i,j)H_l^T(i,j),$$

$$\bar{d}_l((i,j), X, Q(i,j)) \triangleq d_l((i,j), X) + e_l((i,j), Q(i,j)),$$

$$\bar{d}_3(Z(i+1,j), Z(i,j+1)) \triangleq \tau_1 d_1((i+1,j), Z(i+1,j))$$
$$+ \tau_2 d_2((i,j+1), Z(i,j+1)).$$

The procedure for designing the locally optimal filter is presented in the following two theorems.

Theorem 4.2 *Let $\tau > 0$ be a given scalar, $\{\epsilon_l(i,j)\}_{i,j=0}^K$ and $\{\mu_l(i,j)\}_{i,j=0}^K$ ($l = 1, 2$) be four sequences of positive scalars, and assume that there are two sequences of positive definite matrices $\{P(i,j)\}_{i,j=0}^K$ and $\{S(i,j)\}_{i,j=0}^K$ such that*

$$P(i,j) > S(i,j), \tag{4.22}$$

$$P(i+1,j+1) = \bar{d}_1((i+1,j), P(i+1,j), Q(i+1,j))$$
$$+ \bar{d}_2((i,j+1), P(i,j+1), Q(i,j+1))$$
$$+ \bar{d}_3(P(i+1,j), P(i,j+1)), \tag{4.23}$$
$$S(i+1,j+1) = \bar{d}_1((i+1,j), S(i+1,j), Q(i+1,j))$$
$$+ \bar{d}_2((i,j+1), S(i,j+1), Q(i,j+1))$$
$$+ \bar{d}_3(S(i+1,j), S(i,j+1))$$
$$- \tilde{\Omega}_1((i+1,j), S(i+1,j), P(i+1,j))$$
$$- \tilde{\Omega}_2((i,j+1), S(i,j+1), P(i,j+1)), \tag{4.24}$$
$$\epsilon_l^{-1}(i,j)I > M(i,j)P(i,j)M^T(i,j), \tag{4.25}$$
$$\mu_l^{-1}(i,j)I > E(i,j)Q(i,j)E^T(i,j) \tag{4.26}$$

are satisfied with initial conditions

$$P(i,0) > S(i,0) \geq \Xi(i,0), \quad P(0,j) > S(0,j) \geq \Xi(0,j).$$

Then, there is a 2-D quadratic filter (4.4) for system (4.9) with filter parameters given as

$$\hat{A}_l(i,j) = A_l(i,j) + \left[A_l(i,j) - \hat{K}_l(i,j)C(i,j)\bar{\Lambda}_\beta(i,j) \right]$$
$$\times S(i,j)M^T(i,j)\bar{\Omega}_l((i,j), S(i,j))M(i,j), \tag{4.27}$$
$$\hat{K}_l(i,j) = (1+\tau_l)A_l(i,j)\Omega_l((i,j), S(i,j))\bar{\Lambda}_\beta(i,j)$$
$$\times C^T(i,j)\Upsilon_l^{-1}((i,j), S(i,j), P(i,j)) \tag{4.28}$$

such that the estimation error variance is bounded by

$$\tilde{\Xi}(i,j) \leq S(i,j). \tag{4.29}$$

Proof *Based on the definitions of matrix functions $f_l((i,j), X, Q(i,j))$ and $\bar{d}_l((i,j), X, Q(i,j))$, under the condition that matrices X and Y satisfy $Y > 0$, $Y \geq X$, and $\epsilon_l^{-1}(i,j)I - M(i,j)YM^T(i,j) > 0$, the following relationships hold:*

$$f_l((i,j), X, Q(i,j)) \leq f_l((i,j), Y, Q(i,j))$$
$$\leq \bar{d}_l((i,j), Y, Q(i,j)), \quad l = 1,2 \tag{4.30}$$

where (4.26) has been used in the above derivation. In the light of (4.8), (4.10), (4.14), and (4.23), it is evident to see that the expressions of $\Xi(i,j)$ and $P(i,j)$ have similar structures with those of $\bar{\Theta}(i,j)$ and $\Theta(i,j)$, respectively. Recalling (4.30) and adopting an analogous method as shown in the proof of Theorem 4.1, we conclude the statement $\Xi(i,j) \leq P(i,j)$ for all $i,j \in [0\ K]$.

Next, we intend to construct a positive definite matrix $\Theta(i,j)$ satisfying (4.12)–(4.14). The matrix $\Theta(i,j)$ is designed with the following structure:

$$\Theta(i,j) \triangleq \begin{bmatrix} P(i,j) & P(i,j) - S(i,j) \\ P(i,j) - S(i,j) & P(i,j) - S(i,j) \end{bmatrix}. \tag{4.31}$$

It follows from (4.22) that $\Theta(i,j) > 0$. *Besides, conditions (4.25)–(4.26) yield that*

$$\epsilon_l^{-1}(i,j)I - \bar{M}(i,j)\Theta(i,j)\bar{M}^T(i,j)$$
$$= \epsilon_l^{-1}(i,j)I - M(i,j)P(i,j)M^T(i,j) > 0$$

and

$$\mu_l^{-1}(i,j)I - \bar{E}(i,j)\tilde{Q}(i,j)\bar{E}^T(i,j)$$
$$= \mu_l^{-1}(i,j)I - E(i,j)Q(i,j)M^T(i,j) > 0$$

which immediately imply (4.12) and (4.13).

Now, it remains to prove that $\Theta(i,j)$ expressed by (4.31) is a solution to (4.14). Substituting (4.27), (4.28), and (4.31) into the right-hand side of (4.14) and considering the recursive equations (4.23) and (4.24), we have the following equality (4.32) based on some algebraic manipulations:

$$
\begin{aligned}
&(1+\tau)\left[\bar{A}_1(i+1,j)\Psi_1(i+1,j)\bar{A}_1^T(i+1,j)\right.\\
&\left.+\epsilon_1^{-1}(i+1,j)\bar{N}_1(i+1,j)\bar{N}_1^T(i+1,j)\right]\\
&+(1+\tau^{-1})\left[\bar{A}_2(i,j+1)\Psi_2(i,j+1)\bar{A}_2^T(i,j+1)\right.\\
&\left.+\epsilon_2^{-1}(i,j+1)\bar{N}_2(i,j+1)\bar{N}_2^T(i,j+1)\right]\\
&+\bar{B}_1(i+1,j)\left[\tilde{Q}^{-1}(i+1,j)-\mu_1(i+1,j)\bar{E}^T(i+1,j)\bar{E}(i+1,j)\right]^{-1}\\
&\times \bar{B}_1^T(i+1,j)+\bar{B}_2(i,j+1)\\
&\times\left[\tilde{Q}^{-1}(i,j+1)-\mu_2(i,j+1)\bar{E}^T(i,j+1)\bar{E}(i,j+1)\right]^{-1}\bar{B}_2^T(i,j+1)\\
&+\mu_1^{-1}(i+1,j)\bar{H}_1(i+1,j)\bar{H}_1^T(i+1,j)\\
&+\mu_2^{-1}(i,j+1)\bar{H}_2(i,j+1)\bar{H}_2^T(i,j+1)\\
&=\left[\begin{matrix} P(i+1,j+1) & P(i+1,j+1)-S(i+1,j+1)\\ P(i+1,j+1)-S(i+1,j+1) & P(i+1,j+1)-S(i+1,j+1) \end{matrix}\right],
\end{aligned}
$$
$$(4.32)$$

where

$$\Psi_l(i,j) = \left[\Theta^{-1}(i,j) - \epsilon_l(i,j)\bar{M}^T(i,j)\bar{M}(i,j)\right]^{-1}. \qquad (4.33)$$

It is concluded from (4.32) that the constructed $\Theta(i,j)$ *satisfies (4.14). As a result, there is a 2-D quadratic filter for system (4.9), and (4.29) is valid from Corollary 4.1. The proof is complete.*

We now come to the stage of showing that the filter (4.4) with (4.27) and (4.28) is the desired optimal one minimizing the established upper bound.

Theorem 4.3 *Under the same conditions as in Theorem 4.2, the 2-D quadratic filter with parameters designed as (4.27) and (4.28) minimizes the upper bound* $S(i,j)$.

Proof *According to (4.14) and (4.31), it is clear to see that*

$$
\begin{aligned}
S(i+1,j+1) &= [I \ - I]\Theta(i+1,j+1)[I \ - I]^T \\
&= (1+\tau_1)\big[\zeta_1(i+1,j)\Psi_1(i+1,j)\zeta_1^T(i+1,j) \\
&\quad + \epsilon_1^{-1}(i+1,j)N_1(i+1,j)N_1^T(i+1,j)\big] \\
&\quad + (1+\tau_2)\big[\zeta_2(i,j+1)\Psi_2(i,j+1)\zeta_2^T(i,j+1) \\
&\quad + \epsilon_2^{-1}(i,j+1)N_2(i,j+1)N_2^T(i,j+1)\big] \\
&\quad + \hat{K}_1(i+1,j)\tilde{R}(i+1,j)\hat{K}_1^T(i+1,j) \\
&\quad + \hat{K}_2(i,j+1)\tilde{R}(i,j+1)\hat{K}_2^T(i,j+1) \\
&\quad + e_1((i+1,j),Q(i+1,j)) \\
&\quad + e_2((i,j+1)Q((i,j+1)),
\end{aligned}
\tag{4.34}
$$

where $\Psi_l(i,j)$ $(l=1,2)$ and $\tilde{R}(i,j)$ are presented in (4.33) and Definition 4.1, respectively, and

$$
\zeta_l(i,j) = \Big[A_l(i,j) - \hat{K}_l(i,j)C(i,j)\bar{\Lambda}_\beta(i,j) \\
\hat{K}_l(i,j)C(i,j)\bar{\Lambda}_\beta(i,j) - \hat{A}_l(i,j)\Big].
$$

To minimize the upper bound $S(i,j)$ at each instant, we take the first variation to (4.34) with regard to $\hat{A}_1(i+1,j)$, $\hat{A}_2(i,j+1)$, $\hat{K}_1(i+1,j)$, and $\hat{K}_2(i,j+1)$, respectively. Then, the following results are obtained:

$$
\begin{aligned}
S_1(i,j) &\triangleq \partial S(i+1,j+1)/\partial\hat{A}_1(i+1,j) \\
&= (1+\tau_1)\zeta_1(i+1,j)\Psi_1(i+1,j)[0 \ - I]^T \\
&\quad + (1+\tau_1)[0 \ - I]\Psi_1(i+1,j)\zeta_1^T(i+1,j),
\end{aligned}
\tag{4.35}
$$

$$
\begin{aligned}
S_2(i,j) &\triangleq \partial S(i+1,j+1)/\partial\hat{A}_2(i,j+1) \\
&= (1+\tau_2)\zeta_2(i,j+1)\Psi_2(i,j+1)[0 \ - I]^T \\
&\quad + (1+\tau_2)[0 \ - I]\Psi_2(i,j+1)\zeta_2^T(i,j+1),
\end{aligned}
\tag{4.36}
$$

$$
\begin{aligned}
S_3(i,j) &\triangleq \partial S(i+1,j+1)/\partial\hat{K}_1(i+1,j) \\
&= (1+\tau_1)\zeta_1(i+1,j)\Psi_1(i+1,j)[-I\ I]^T\bar{\Lambda}_\beta^T(i+1,j)C^T(i+1,j) \\
&\quad + (1+\tau_1)C(i+1,j)\bar{\Lambda}_\beta(i+1,j)[-I\ I]\Psi_1(i+1,j)\zeta_1^T(i+1,j) \\
&\quad + \hat{K}_1(i+1,j)\tilde{R}(i+1,j) + \tilde{R}^T(i+1,j)\hat{K}_1^T(i+1,j),
\end{aligned}
\tag{4.37}
$$

$$
\begin{aligned}
S_4(i,j) &\triangleq \partial S(i+1,j+1)/\partial\hat{K}_2(i,j+1) \\
&= (1+\tau_2)\zeta_2(i,j+1)\Psi_2(i,j+1)[-I\ I]^T\bar{\Lambda}_\beta^T(i,j+1)C^T(i,j+1) \\
&\quad + (1+\tau_2)C(i,j+1)\bar{\Lambda}_\beta(i,j+1)[-I\ I]\Psi_2(i,j+1)\zeta_2^T(i,j+1) \\
&\quad + \hat{K}_2(i,j+1)\tilde{R}(i,j+1) + \tilde{R}^T(i,j+1)\hat{K}_2^T(i,j+1).
\end{aligned}
\tag{4.38}
$$

Some tedious yet routine algebraic computations yield $S_k(i,j) = 0$ ($k \in$ [1 4]) if the filter parameters are determined by (4.27) and (4.28).

On the other hand, let us denote matrices $\Psi_l(i,j)$ in (4.33) as

$$\Psi_l(i,j) \triangleq \begin{bmatrix} \Psi_l^{(11)}(i,j) & \Psi_l^{(12)}(i,j) \\ (\Psi_l^{(12)}(i,j))^T & \Psi_l^{(22)}(i,j) \end{bmatrix} > 0. \qquad (4.39)$$

It is not difficult to calculate that the second variations of (4.34) with regard to $\hat{A}_1(i+1,j)$, $\hat{A}_2(i,j+1)$, $\hat{K}_1(i+1,j)$, and $\hat{K}_2(i,j+1)$ are all positive definite, namely,

$$\partial^2 S(i+1,j+1)/\partial \hat{A}_1^2(i+1,j) = 2(1+\tau_1)\Psi_1^{(22)}(i+1,j) > 0, \qquad (4.40)$$

$$\partial^2 S(i+1,j+1)/\partial \hat{A}_2^2(i,j+1) = 2(1+\tau_2)\Psi_2^{(22)}(i,j+1) > 0, \qquad (4.41)$$

$$\partial^2 S(i+1,j+1)/\partial \hat{K}_1^2(i+1,j)$$
$$= 2(1+\tau_1)C(i+1,j)\bar{\Lambda}_\beta(i+1,j)[-I\ I]\Psi_1(i+1,j)$$
$$\times [-I\ I]^T \bar{\Lambda}_\beta^T(i+1,j)C^T(i+1,j) + 2\tilde{R}^T(i+1,j) > 0, \qquad (4.42)$$

$$\partial^2 S(i+1,j+1)/\partial \hat{K}_2^2(i,j+1)$$
$$= 2(1+\tau_2)C(i,j+1)\bar{\Lambda}_\beta(i,j+1)[-I\ I]\Psi_2(i,j+1)$$
$$\times [-I\ I]^T \bar{\Lambda}_\beta^T(i,j+1)C^T(i,j+1) + 2\tilde{R}^T(i,j+1) > 0. \qquad (4.43)$$

Consequently, the filter parameters given as (4.27) and (4.28) are optimal, and thus the theorem is proved.

Remark 4.5 *From Theorems 4.2–4.3, the existence of positive definite matrix sequences $P(i,j)$ and $S(i,j)$ subject to (4.22)–(4.26) ensures a filter (4.4) achieving the filter design objectives with gain parameters determined by (4.27) and (4.28), namely, an upper bound is constructed and subsequently minimized for the estimation error variance. Moreover, the expected filter and the minimal bound are presented in terms of recursive Riccati-like equations, which are suitable for online applications.*

Remark 4.6 *It is worth mentioning that a stochastic Kronecker delta function is adopted in this chapter to describe the incomplete measurements comprising the randomly varying sensor delays and the missing measurements, which would influence the filtering performance. Intuitively, a large arrival probability of the current measurement indicates that the sensor works well, and the increase of probabilities $\bar{\beta}_2(i,j)$ and $\bar{\beta}_3(i,j)$ represents a higher occurrence rate of sensor delays in a statistical sense. On the other hand, the sensor working condition is recognized to be worse as the value of $\bar{\beta}_4(i,j)$ increases. Therefore, the filter performance is expected to be better with more available measurements or less rate of the missing measurements.*

Remark 4.7 *It is of practical interest and importance to keep the upper bound* $S(i,j)$ *as tight as possible under the condition that (4.22)–(4.26) are solvable, where the scaling parameters* $\epsilon_l(i,j)$ *and* $\mu_l(i,j)$ $(l = 1,2)$ *are introduced for establishing the Riccati-like equations (4.23) and (4.24). Here, the solvability of these equations is dependent on the choice of* $\epsilon_l(i,j)$ *and* $\mu_l(i,j)$, *which would have a great effect on the existence/determination of the minimal upper bound* $S(i,j)$. *Although the feasibility and convergence properties of the Riccati-like equations have been discussed in [263] for the 1-D robust Kalman filtering problem, it would be quite challenging to explore the issue of choosing appropriate parameters* $\epsilon_l(i,j)$ *and* $\mu_l(i,j)$ *in the 2-D filter design process, which deserves special focus in the future research.*

4.4   Numerical Example

In this section, a simulation example is presented to demonstrate the effectiveness of the filter design scheme.

Consider the 2-D system (4.1) over a finite horizon $i,j \in [0 \; 50]$ with the following parameters:

$$\vec{A}_1(i,j) = \begin{bmatrix} 0.55 & 0.2\sin(3i) \\ 0.1 & -0.4 + 0.1\cos(j) \end{bmatrix},$$

$$\vec{A}_2(i,j) = \begin{bmatrix} -0.2 + 0.15\cos(2j) & 0 \\ 0.1\sin(i) & 0.4 \end{bmatrix},$$

$$\vec{B}_1(i,j) = \begin{bmatrix} 0.12 - 0.1e^{-2i} \\ 0.35 \end{bmatrix},$$

$$\vec{B}_2(i,j) = \begin{bmatrix} 0.28 \\ 0.15 + 0.05\sin(-j) \end{bmatrix},$$

$$\vec{N}_1(i,j) = \begin{bmatrix} -0.25 \\ 0.16 \end{bmatrix}, \quad \vec{N}_2(i,j) = \begin{bmatrix} 0.5 \\ 0.1 \end{bmatrix},$$

$$\vec{H}_1(i,j) = \begin{bmatrix} 0.15 \\ 0.28 \end{bmatrix}, \quad \vec{H}_2(i,j) = \begin{bmatrix} 0.2 \\ -0.35 \end{bmatrix},$$

$$F(i,j) = \sin(0.6i)\cos(0.1j), \quad M(i,j) = [0.1 \; 0.2],$$

$$E(i,j) = -0.15, \quad \vec{C}(i,j) = [0.4 + 0.1\sin(i) \; 0.1].$$

The random variables $w(i,j)$ and $v(i,j)$ are assumed to be uncorrelated Gaussian white noises with mean zero and variances $Q(i,j) = 0.5$ and $R(i,j) = 0.16$, respectively. The initial conditions associated with the addressed system are chosen as $\vec{u}_1(i) = \vec{u}_2(j) = [0 \; 0]^T$ for $i,j \in [0 \; 50]$, and $\vec{\Xi}(i,0) = 0.12I$ and $\vec{\Xi}(0,j) = 0.1I$ for $i,j \in [1 \; 50]$ which result in

$$\Xi(1,0) = \text{diag}\{0.12, 0.12, 0, 0, 0, 0\},$$

$$\Xi(0,1) = \mathrm{diag}\{0.1, 0.1, 0, 0, 0, 0\},$$
$$\Xi(i,0) = \mathrm{diag}\{0.12, 0.12, 0, 0, 0.12, 0.12\},$$
$$\Xi(0,j) = \mathrm{diag}\{0.1, 0.1, 0.1, 0.1, 0, 0\}, \quad i,j \in [2\ 50].$$

Besides, we set $\tau = 1$, $\epsilon_l(i,j) = \mu_l(i,j) = 10$ $(l = 1,2)$ for $i,j \in [0\ 50]$, $S(0,0) = 0.11I$, $P(0,0) = 0.125I$, $S(i,0) = 0.12I$, $S(0,j) = 0.1I$, $P(i,0) = 0.13I$, and $P(0,j) = 0.12I$ for $i,j \in [1\ 50]$.

For notational simplicity, the estimation error and the actual estimation error variance are respectively denoted as

$$x(i,j) - \hat{x}(i,j) \triangleq e(i,j),$$
$$\mathbb{E}\{e(i,j)e(i,j)^T\} \triangleq \tilde{\Xi}(i,j).$$

To show the usefulness of the designed 2-D filter, the following case is considered firstly. **Case 1:** The probability distribution of stochastic variable $\beta(i,j)$ is given as $\bar{\beta}_1(i,j) = 0.65$, $\bar{\beta}_2(i,j) = \bar{\beta}_3(i,j) = 0.15$, and $\bar{\beta}_4(i,j) = 0.05$. Accordingly, scalars $\hat{\beta}_{kl}(i,j)$ with $\hat{\beta}_{kl}(i,j) = \hat{\beta}_{lk}(i,j)$ for $k,l \in [1\ 3]$ are easily calculated as $\hat{\beta}_{11}(i,j) = 0.2275$, $\hat{\beta}_{22}(i,j) = \hat{\beta}_{33}(i,j) = 0.1275$, $\hat{\beta}_{12}(i,j) = \hat{\beta}_{13}(i,j) = -0.0975$, and $\hat{\beta}_{23}(i,j) = -0.0225$. From Theorem 4.2, matrices $P(i,j)$ and $S(i,j)$ as well as filter parameters $\hat{A}_l(i,j)$ and $\hat{K}_l(i,j)$ can be calculated by iteratively solving (4.23), (4.24), (4.27), and (4.28), which are not presented here for space consideration. On the other hand, matrices $S(i,j)$ and $S(i,j) - \tilde{\Xi}(i,j)$ are taken into account to characterize the filtering performance. For space consideration, Fig. 4.1 depicts the trajectory of $e^{(1)}(i,j)$ representing the first element of $e(i,j)$, and Figs. 4.2–4.3 plot the variance upper bound $S^{(11)}(i,j)$ and the difference between $S^{(11)}(i,j)$ and the real estimation error variance $\tilde{\Xi}^{(11)}(i,j)$, respectively, which demonstrate that the designed 2-D filter performs well and achieves the expected objectives.

To illustrate the influence of incomplete measurements on the filtering performance, three different cases are also taken into consideration.

Case 2: In this case, the arrival probabilities of measurements are reset as $\bar{\beta}_1(i,j) = 0.3$, $\bar{\beta}_2(i,j) = \bar{\beta}_3(i,j) = 0.15$, and $\bar{\beta}_4(i,j) = 0.4$ without modifying the other parameters given in the above case. It is easy to compute that $\hat{\beta}_{11}(i,j) = 0.21$, $\hat{\beta}_{22}(i,j) = \hat{\beta}_{33}(i,j) = 0.1275$, $\hat{\beta}_{12}(i,j) = \hat{\beta}_{13}(i,j) = -0.045$, and $\hat{\beta}_{23}(i,j) = -0.0225$. The simulation results are shown in Figs. 4.4–4.5.

Case 3: The probabilities are further selected as $\bar{\beta}_1(i,j) = 0$ and $\bar{\beta}_4(i,j) = 0.7$, while the other parameters remain the same as in Case 1. One has $\hat{\beta}_{11}(i,j) = \hat{\beta}_{12}(i,j) = \hat{\beta}_{13}(i,j) = 0$ immediately. Figures 4.6–4.7 depict the simulation results of this case.

Case 4: In the last case, by choosing the probabilities $\bar{\beta}_k(i,j) = 0$ for $k \in [1\ 3]$ and $\bar{\beta}_4(i,j) = 1$, while remaining the rest parameters the same as above, one has $\Pi_\beta(i,j) = 0$. The simulation results are shown in Figs. 4.8–4.9.

It can be seen from the above cases that the locally minimal upper bound $S^{(11)}(i,j)$ for $\tilde{\Xi}^{(11)}(i,j)$ and the corresponding difference between them are tighter with the increased arrival probabilities $\sum_{k=1}^{3} \bar{\beta}_k(i,j)$ for the useful

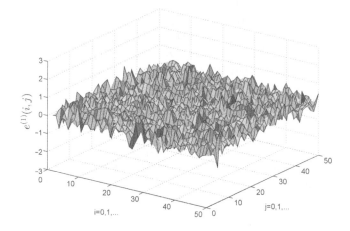

FIGURE 4.1: Estimation error $e^{(1)}(i, j)$ for Case 1.

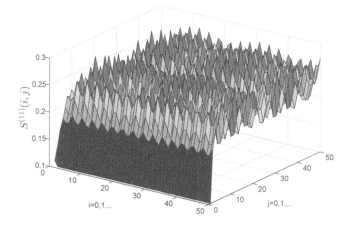

FIGURE 4.2: Evolution of the upper bound $S^{(11)}(i, j)$ for Case 1.

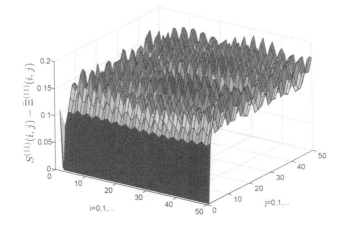

FIGURE 4.3: Difference between $S^{(11)}(i,j)$ and $\tilde{\Xi}^{(11)}(i,j)$ for Case 1.

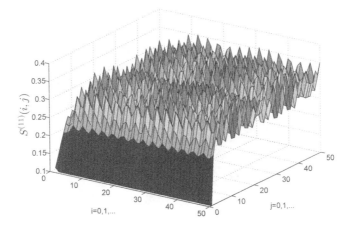

FIGURE 4.4: Evolution of the upper bound $S^{(11)}(i,j)$ for Case 2.

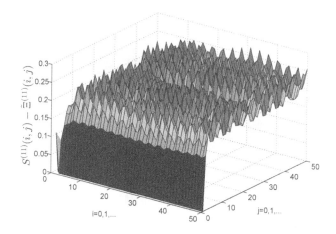

FIGURE 4.5: Difference between $S^{(11)}(i,j)$ and $\tilde{\Xi}^{(11)}(i,j)$ for Case 2.

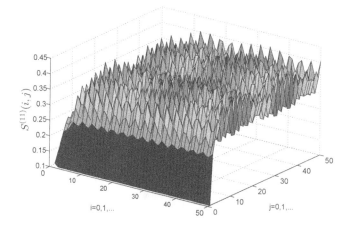

FIGURE 4.6: Evolution of the upper bound $S^{(11)}(i,j)$ for Case 3.

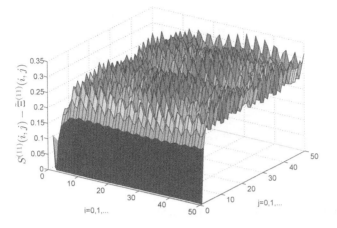

FIGURE 4.7: Difference between $S^{(11)}(i,j)$ and $\tilde{\Xi}^{(11)}(i,j)$ for Case 3.

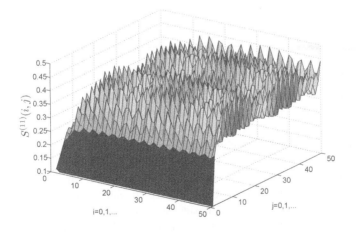

FIGURE 4.8: Evolution of the upper bound $S^{(11)}(i,j)$ for Case 4.

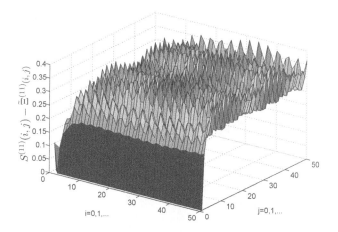

FIGURE 4.9: Difference between $S^{(11)}(i,j)$ and $\tilde{\Xi}^{(11)}(i,j)$ for Case 4.

information that includes both the current and delayed measurements. Particularly, by comparing Case 1 with Cases 2–3, one concludes that for the given arrival probabilities $\bar{\beta}_k(i,j)$ ($k = 2,3$) of possible delayed measurements, the minimum upper bound $S^{(11)}(i,j)$ and the difference $S^{(11)}(i,j) - \tilde{\Xi}^{(11)}(i,j)$ become less compact when decreasing $\bar{\beta}_1(i,j)$ and increasing $\bar{\beta}_4(i,j)$. In comparison with Cases 1–4, one can see that the filtering performance is in its worst case when only noises are received.

4.5   Summary

This chapter has investigated the robust filtering problem for 2-D shift-varying uncertain systems with randomly varying sensor delays over a finite horizon. By adopting a stochastic Kronecker delta function, a unified form has been proposed to describe the received measurements containing time delays and missing measurements, both of which occur randomly with certain probability distributions. Owing to the presence of the random and uncertain nature in the tackled system, it is impossible to derive the actual estimation error variance. Hence, according to mathematical induction, an upper bound has been constructed for the estimation error variance which is locally minimized with a proper 2-D filter. The expected gain parameters have been designed by solving two sequences of recursive Riccati-like equations which can be calculated iteratively. The effectiveness of the proposed filter has also been demonstrated through a simulation example.

5

Recursive Filtering for 2-D Systems with
Missing Measurements subject to Uncertain
Probabilities

For the filtering problems, missing measurements or packet dropouts are frequently encountered due to network constraints, for example, the bandwidth. Moreover, missing measurements are likely to emerge in probabilistic fashion because of the randomly fluctuated environment. Such a phenomenon has been viewed as a key factor affecting the system dynamics and may seriously degrade the system performance if not carefully tackled. Till now, the recursive filtering problem for 2-D systems subject to randomly missing measurements has attracted some initial research attention.

The arrival probabilities of missing measurements are usually assumed to be deterministic in most of the available literature [109, 140, 211]. Nevertheless, it would be more realistic to suppose that the occurrence rates of packet dropouts are uncertain because the exact probabilities are often inaccessible. As a result, the concept of missing measurements with uncertain probabilities has been proposed and considered in [65, 78, 216]. In particular, the gain-scheduled filtering has been designed in [216] for systems with missing measurements, where the occurrence rates are uncertain but in given measured ranges. Recently, the recursive filtering problem has been investigated in [78] for 1-D systems with multiple packet losses suffering from uncertain missing probabilities. However, the corresponding issue has not been investigated for 2-D systems with uncertain missing measurements, let alone the case where shift-varying parameters and variance constraints are simultaneously concerned.

In response to the above discussions, we intend to deal with the recursive filtering problem for 2-D shift-varying systems with missing measurements. The main contributions of this chapter lie in the following two aspects: (1) missing measurements under consideration are governed by Bernoulli distributed white sequences with uncertain probabilities rather than deterministic ones, and (2) by applying the stochastic analysis and induction method, an upper bound is developed for the filtering error variance and the locally optimal one is then obtained with the desired filter gain parameters.

The rest of this chapter is outlined as follows. Section 5.1 formulates the recursive filtering problem to be addressed. The filter design strategy is established in Section 5.2 for the considered system, where the minimal upper bound on the filtering error variance is obtained. Section 5.3 provides a numerical example to show the effectiveness of the main results. Conclusions are drawn in Section 5.4.

5.1   Problem Formulation

Consider a 2-D shift-varying system of the following form:

$$
\begin{aligned}
x(i+1,j+1) &= A_1(i+1,j)x(i+1,j) + A_2(i,j+1)x(i,j+1) \\
&\quad + B_1(i+1,j)w(i+1,j) + B_2(i,j+1)w(i,j+1), \quad \text{(5.1a)} \\
y(i,j) &= \Lambda(i,j)C(i,j)x(i,j) + v(i,j), \quad\quad\quad\quad\quad\quad \text{(5.1b)}
\end{aligned}
$$

where i,j are defined on a finite horizon $[0\ N]$ with N being a fixed positive integer, and $x(i,j) \in \mathbb{R}^n$ and $y(i,j) \in \mathbb{R}^m$ are the state vector and the measurement output, respectively, $w(i,j) \in \mathbb{R}^p$ and $v(i,j) \in \mathbb{R}^m$ are zero-mean Gaussian white-noise sequences with respective covariances $Q(i,j)$ and $R(i,j) > 0$. $A_l(i,j)$, $B_l(i,j)$ $(l = 1,2)$, and $C(i,j)$ are known and shift-varying matrices, while $\Lambda(i,j) \triangleq \mathrm{diag}\{\lambda_1(i,j),\lambda_2(i,j),\ldots,\lambda_m(i,j)\}$ is a random matrix with $\lambda_s(i,j)$ $(s \in [1\ m])$ being m mutually uncorrelated stochastic variables. It is assumed that variable $\lambda_s(i,j)$ for any $s \in [1\ m]$ satisfies the following Bernoulli distribution taking values on 1 or 0:

$$
\mathrm{Prob}\{\lambda_s(i,j) = 1\} = \bar{\lambda}_s + \Delta\lambda_s, \quad\quad\quad\quad \text{(5.2)}
$$
$$
\mathrm{Prob}\{\lambda_s(i,j) = 0\} = 1 - \bar{\lambda}_s - \Delta\lambda_s, \quad\quad \text{(5.3)}
$$

where $\bar{\lambda}_s + \Delta\lambda_s \in [0\ 1]$ is referred to as the arrival probability of the s-th element of the missing measurements, in which $\bar{\lambda}_s$ is a known scalar and $\Delta\lambda_s$ is an unknown one introduced to describe the uncertain probability. The initial conditions for (5.1) are given as $\mathbb{E}\{x(i,0)\} = u_1(i)$ and $\mathbb{E}\{x(0,j)\} = u_2(j)$ $(i,j \in [0\ N])$, where $u_1(i)$ and $u_2(j)$ are known vectors satisfying $u_1(0) = u_2(0)$.

Denote

$$
\begin{aligned}
\bar{\Lambda} &\triangleq \mathrm{diag}_{1 \le s \le m}\{\bar{\lambda}_s\}, \\
\Delta\Lambda &\triangleq \mathrm{diag}_{1 \le s \le m}\{\Delta\lambda_s\}, \\
\tilde{\Lambda}(i,j) &\triangleq \mathrm{diag}_{1 \le s \le m}\{\tilde{\lambda}_s(i,j)\},
\end{aligned}
$$

where $\tilde{\lambda}_s(i,j) = \lambda_s(i,j) - \bar{\lambda}_s - \Delta\lambda_s$. Then, one has

$$
\begin{aligned}
\Lambda(i,j) &= \bar{\Lambda} + \Delta\Lambda + \tilde{\Lambda}(i,j), \\
\mathbb{E}\{\Lambda(i,j)\} &= \bar{\Lambda} + \Delta\Lambda.
\end{aligned}
$$

Throughout this chapter, for $i, j \in [0\ N]$, initial conditions $x(i, 0)$ and $x(0, j)$, random variables $w(i, j)$, $v(i, j)$, and $\lambda_s(i, j)$ $(s \in [1\ m])$ are assumed to be uncorrelated with each other.

For system (5.1), a recursive filter is adopted as follows:

$$\hat{x}_p(i, j) = A_1(i, j-1)\hat{x}_u(i, j-1) + A_2(i-1, j)\hat{x}_u(i-1, j), \tag{5.4a}$$

$$\hat{x}_u(i, j) = \hat{x}_p(i, j) + K(i, j)\left[y(i, j) - \bar{\Lambda}C(i, j)\hat{x}_p(i, j)\right], \tag{5.4b}$$

where i and j belong to $\in [1\ N]$, $\hat{x}_p(i, j)$ is the one-step prediction, $\hat{x}_u(i, j)$ is the updated estimate of state $x(i, j)$, and $K(i, j)$ is the filter parameter to be determined. Initial conditions related to (5.4) are set as $\hat{x}_u(i, 0) = u_1(i)$ and $\hat{x}_u(0, j) = u_2(j)$ for $i, j \in [0\ N]$.

Remark 5.1 *The adopted filter has a two-step structure, i.e., the prediction (5.4a) and the estimation (5.4b), where the innovation information contributes to update estimate at each instant. Such a filter has a similar framework of the conventional Kalman filter for 1-D systems, and hence is referred to as a 2-D Kalman filter.*

Defining the prediction error as $\tilde{x}_p(i, j) \triangleq x(i, j) - \hat{x}_p(i, j)$ and the filtering error as $\tilde{x}_u(i, j) \triangleq x(i, j) - \hat{x}_u(i, j)$, we obtain from from (5.1) and (5.4) that

$$\begin{aligned}\tilde{x}_p(i, j) = &A_1(i, j-1)\tilde{x}_u(i, j-1) + A_2(i-1, j)\tilde{x}_u(i-1, j) \\ &+ B_1(i, j-1)w(i, j-1) + B_2(i-1, j)w(i-1, j),\end{aligned} \tag{5.5a}$$

$$\begin{aligned}\tilde{x}_u(i, j) = &\left(I - K(i, j)\bar{\Lambda}C(i, j)\right)\tilde{x}_p(i, j) - K(i, j)v(i, j) \\ &- K(i, j)\left(\Delta\Lambda + \tilde{\Lambda}(i, j)\right)C(i, j)x(i, j).\end{aligned} \tag{5.5b}$$

The prediction and filtering error variances in a general sense are respectively denoted as

$$P_p(i, j) \triangleq \mathbb{E}\{\tilde{x}_p(i, j)\tilde{x}_p^T(i, j)\},$$
$$P_u(i, j) \triangleq \mathbb{E}\{\tilde{x}_u(i, j)\tilde{x}_u^T(i, j)\}.$$

The primary aim of this chapter is to design a filter of form (5.4) such that an upper bound on the filtering error variance exists and can be minimized with a proper filter parameter at each instant.

Remark 5.2 *Owing to the existence of uncertain missing measurements, the filtering error variance $P_u(i, j)$ is impossible to be calculated accurately, not to mention the design of the gain parameter $K(i, j)$ to optimize $P_u(i, j)$. As such, a tight upper bound of $P_u(i, j)$ is recognized as an alternative way to design the filter as well as reflect the filtering performance.*

5.2   Recursive Filter Design

In this section, we deal with the 2-D filter design scheme for system (5.1). After giving the recursion of the error variances, an upper bound is obtained for the filtering error variance. Subsequently, the filter parameter is properly selected to minimize the developed upper bound at each step.

The following theorems provide the dynamics of the prediction and estimation error variances, respectively.

Theorem 5.1 *For system (5.5), the evolution of the prediction error variance* $P_p(i,j)$ *satisfies the following equation:*

$$
\begin{aligned}
P_p(i,j) = {} & A_1(i,j-1)P_u(i,j-1)A_1^T(i,j-1) \\
& + A_2(i-1,j)P_u(i-1,j)A_2^T(i-1,j) \\
& + A_1(i,j-1)\mathbb{E}\left\{\tilde{x}_u(i,j-1)\tilde{x}_u^T(i-1,j)\right\}A_2^T(i-1,j) \\
& + A_2(i-1,j)\mathbb{E}\left\{\tilde{x}_u(i-1,j)\tilde{x}_u^T(i,j-1)\right\}A_1^T(i,j-1) \\
& + B_1(i,j-1)Q(i,j-1)B_1^T(i,j-1) \\
& + B_2(i-1,j)Q(i-1,j)B_2^T(i-1,j), \qquad i,j \in [1 \ N]. \qquad (5.6)
\end{aligned}
$$

Proof *The proof is confirmed directly from (5.5a) where the details are omitted for brevity.*

Theorem 5.2 *For system (5.5), the evolution of the filtering error variance* $P_u(i,j)$ *is governed by*

$$
\begin{aligned}
P_u(i,j) = {} & \left[I - K(i,j)\bar{\Lambda}C(i,j)\right]P_p(i,j)\left[I - K(i,j)\bar{\Lambda}C(i,j)\right]^T \\
& + K(i,j)\Delta\Lambda C(i,j)\mathbb{E}\left\{x(i,j)x^T(i,j)\right\}\left(K(i,j)\Delta\Lambda C(i,j)\right)^T \\
& + K(i,j)\mathbb{E}\left\{\tilde{\Lambda}(i,j)C(i,j)x(i,j)x^T(i,j)C^T(i,j)\tilde{\Lambda}^T(i,j)\right\}K^T(i,j) \\
& - \left[I - K(i,j)\bar{\Lambda}C(i,j)\right]\mathbb{E}\left\{\tilde{x}_p(i,j)x^T(i,j)\right\}\left(K(i,j)\Delta\Lambda C(i,j)\right)^T \\
& - K(i,j)\Delta\Lambda C(i,j)\mathbb{E}\left\{x(i,j)\tilde{x}_p^T(i,j)\right\}\left[I - K(i,j)\bar{\Lambda}C(i,j)\right]^T \\
& + K(i,j)R(i,j)K^T(i,j), \qquad i,j \in [1 \ N]. \qquad (5.7)
\end{aligned}
$$

Proof *It is easy to draw the conclusion of this theorem from (5.5b), and thus the proof is omitted here for brevity.*

Noting that the uncertainty induced from the missing measurements is involved in (5.7), hence it is impossible to obtain the analytical solution of the error variance $P_u(i,j)$ and the optimal gain parameter $K(i,j)$ that minimizes $P_u(i,j)$. Therefore, we turn to find a tight upper bound of $P_u(i,j)$ as an alternative strategy to determine the filter gain with which the minimal bound can be achieved.

Theorem 5.3 *Let* μ, γ, *and* ϵ *be given positive scalars. If there are two sets of positive definite matrices* $\{M_p(i,j)\}$ *and* $\{M_u(i,j)\}$ *such that*

$$
\begin{aligned}
M_p(i,j) &= (1+\mu)A_1(i,j-1)M_u(i,j-1)A_1^T(i,j-1) \\
&\quad + (1+\mu^{-1})A_2(i-1,j)M_u(i-1,j)A_2^T(i-1,j) \\
&\quad + B_1(i,j-1)Q(i,j-1)B_1^T(i,j-1) \\
&\quad + B_2(i-1,j)Q(i-1,j)B_2^T(i-1,j), \quad\quad\quad (5.8)
\end{aligned}
$$

$$
\begin{aligned}
M_u(i,j) &= (1+\gamma)[I - K(i,j)\bar{A}C(i,j)]M_p(i,j)[I - K(i,j)\bar{A}C(i,j)]^T \\
&\quad + K(i,j)\big[(1+\gamma^{-1})\sigma^2 \mathrm{tr}\{C(i,j)\Pi(i,j)C^T(i,j)\}I \\
&\quad + \frac{1}{4}I \circ (C(i,j)\Pi(i,j)C^T(i,j)) + R(i,j)\big]K^T(i,j) \quad (5.9)
\end{aligned}
$$

with initial constraints

$$
M_u(i,0) = P_u(i,0), \quad M_u(0,j) = P_u(0,j), \quad i,j \in [1\ N]
$$

are satisfied, where

$$
\sigma \triangleq \max_{1\le s\le m}\left\{\bar{\lambda}_s, 1-\bar{\lambda}_s\right\},
$$
$$
\Pi(i,j) \triangleq (1+\epsilon)M_p(i,j) + (1+\epsilon^{-1})\hat{x}_p(i,j)\hat{x}_p^T(i,j)
$$

then, matrices $M_p(i,j)$ *and* $M_u(i,j)$ *are upper bounds for* $P_p(i,j)$ *and* $P_u(i,j)$ *($i,j \in [1\ N]$), respectively, i.e.,*

$$
P_p(i,j) \le M_p(i,j), \quad P_u(i,j) \le M_u(i,j). \quad\quad (5.10)
$$

Proof *The proof of this theorem is concluded through the inductive approach.*

At the beginning, it follows from Lemma 2.1 and Lemma 3.1 that

$$
\begin{aligned}
\mathbb{E}\{x(i,j)x^T(i,j)\} &\le (1+\epsilon)P_p(i,j) + (1+\epsilon^{-1})\hat{x}_p(i,j)\hat{x}_p^T(i,j) \\
&\triangleq \bar{\Pi}(i,j), \quad\quad\quad (5.11)
\end{aligned}
$$
$$
\mathbb{E}\{\tilde{\Lambda}(i,j)X\tilde{\Lambda}^T(i,j)\} \le \frac{1}{4}I \circ X, \quad\quad (5.12)
$$

where X is a given matrix. Besides, recalling (5.2) and (5.3), one has $-\bar{\lambda}_s \le \Delta\lambda_s \le 1-\bar{\lambda}_s$, which means $\Delta\lambda_s \le \sigma$ for all $s \in [1\ m]$ and thus

$$
\Delta\Lambda \le \sigma I. \quad\quad (5.13)
$$

Then, by resorting to Lemma 3.1 once again, we have

$$
\begin{aligned}
P_p(i,j) &\le (1+\mu)A_1(i,j-1)P_u(i,j-1)A_1^T(i,j-1) \\
&\quad + (1+\mu^{-1})A_2(i-1,j)P_u(i-1,j)A_2^T(i-1,j)
\end{aligned}
$$

$$+ B_1(i, j-1)Q(i, j-1)B_1^T(i, j-1)$$
$$+ B_2(i-1, j)Q(i-1, j)B_2^T(i-1, j). \qquad (5.14)$$

Substituting (5.11)–(5.13) into (5.7) yields

$$P_u(i, j) \le (1+\gamma)\left[I - K(i, j)\bar{\Lambda}C(i, j)\right]P_p(i, j)\left[I - K(i, j)\bar{\Lambda}C(i, j)\right]^T$$
$$+ K(i, j)\left[(1+\gamma^{-1})\sigma^2\mathrm{tr}\left\{C(i, j)\bar{\Pi}(i, j)C^T(i, j)\right\}I\right.$$
$$\left. + \frac{1}{4}I \circ (C(i, j)\bar{\Pi}(i, j)C^T(i, j)) + R(i, j)\right]K^T(i, j). \qquad (5.15)$$

According to (5.8), (5.9), (5.14), and (5.15), it is clear to see

$$P_p(1, 1) - M_p(1, 1)$$
$$\le (1+\mu)A_1(1, 0)\left[P_u(1, 0) - M_u(1, 0)\right]A_1^T(1, 0)$$
$$+ (1+\mu^{-1})A_2(0, 1)\left[P_u(0, 1) - M_u(0, 1)\right]A_2^T(0, 1)$$
$$= 0 \qquad (5.16)$$

which signifies

$$P_u(1, 1) - M_u(1, 1)$$
$$\le (1+\gamma)\left[I - K(1, 1)\bar{\Lambda}C(1, 1)\right]\left[P_p(1, 1) - M_p(1, 1)\right]\left[I - K(1, 1)\bar{\Lambda}C(1, 1)\right]^T$$
$$+ K(1, 1)\left\{(1+\gamma^{-1})\sigma^2\mathrm{tr}\left\{C(1, 1)\bar{\Pi}(1, 1)C^T(1, 1)\right\}I\right.$$
$$- (1+\gamma^{-1})\sigma^2\mathrm{tr}\left\{C(1, 1)\Pi(1, 1)C^T(1, 1)\right\}I$$
$$\left. + \frac{1}{4}I \circ \left[C(1, 1)(\bar{\Pi}(1, 1) - \Pi(1, 1))C^T(1, 1)\right]\right\}K^T(1, 1)$$
$$\le 0. \qquad (5.17)$$

Suppose that (5.10) holds for $i, j \in [1 \ N-1]$ and $i+j=k$ with certain constant $k \in [2 \ 2N-1]$. Then, for i, j belonging to $[1 \ N]$ and satisfying $i+j=k+1$, it is not difficult to derive that

$$P_p(i, j) - M_p(i, j)$$
$$\le (1+\mu)A_1(i, j-1)[P_u(i, j-1) - M_u(i, j-1)]A_1^T(i, j-1)$$
$$+ (1+\mu^{-1})A_2(i-1, j)[P_u(i-1, j) - M_u(i-1, j)]A_2^T(i-1, j)$$
$$\le 0 \qquad (5.18)$$

which also means

$$P_u(i, j) - M_u(i, j)$$
$$\le (1+\gamma)\left[I - K(i, j)\bar{\Lambda}C(i, j)\right][P_p(i, j) - M_p(i, j)]\left[I - K(i, j)\bar{\Lambda}C(i, j)\right]^T$$
$$+ K(i, j)\left\{(1+\gamma^{-1})\sigma^2\mathrm{tr}\left\{C(i, j)\bar{\Pi}(i, j)C^T(i, j)\right\}I\right.$$

$$- (1 + \gamma^{-1})\sigma^2 \text{tr} \left\{ C(i,j)\Pi(i,j)C^T(i,j) \right\} I$$
$$+ \frac{1}{4} I \circ \left[C(i,j)(\bar{\Pi}(i,j) - \Pi(i,j))C^T(i,j) \right] \Big\} K^T(i,j)$$
$$\leq 0. \tag{5.19}$$

The proof is hence concluded by the induction process.

Theorem 5.4 *With the same notations as defined in Theorem 5.3, the minimal upper bound $M_u(i,j)$ has the following form*

$$M_u(i,j) = (1 + \gamma)M_p(i,j) - (1 + \gamma)^2 M_p(i,j)C^T(i,j)$$
$$\times \bar{\Lambda}^T \Phi^{-1}(i,j)\bar{\Lambda}C(i,j)M_p(i,j) \tag{5.20}$$

at each step by choosing the filter gain as

$$K(i,j) = (1 + \gamma)M_p(i,j)C^T(i,j)\bar{\Lambda}^T \Phi^{-1}(i,j), \tag{5.21}$$

where

$$\Phi(i,j) = (1 + \gamma)\bar{\Lambda}C(i,j)M_p(i,j)C^T(i,j)\bar{\Lambda}^T$$
$$+ (1 + \gamma^{-1})\sigma^2 \text{tr} \left\{ C(i,j)\Pi(i,j)C^T(i,j) \right\} I$$
$$+ \frac{1}{4} I \circ \left(C(i,j)\Pi(i,j)C^T(i,j) \right) + R(i,j).$$

Proof *By using the completing-the-square approach, we rewrite (5.9) as*

$$M_u(i,j) = (1 + \gamma)M_p(i,j) + K(i,j)\Phi(i,j)K^T(i,j)$$
$$- (1 + \gamma)K(i,j)\bar{\Lambda}C(i,j)M_p(i,j)$$
$$- (1 + \gamma)M_p(i,j)\left(K(i,j)\bar{\Lambda}C(i,j) \right)^T$$
$$= (1 + \gamma)M_p(i,j) + \left[K(i,j) - (1 + \gamma)M_p(i,j)C^T(i,j)\bar{\Lambda}^T \Phi^{-1}(i,j) \right]$$
$$\times \Phi(i,j) \left[K(i,j) - (1 + \gamma)M_p(i,j)C^T(i,j)\bar{\Lambda}^T \Phi^{-1}(i,j) \right]^T$$
$$- (1 + \gamma)^2 M_p(i,j)C^T(i,j)\bar{\Lambda}^T \Phi^{-1}(i,j)\bar{\Lambda}C(i,j)M_p(i,j).$$

Obviously, the filter parameter should be determined as (5.21) in order to minimize the upper bound $M_u(i,j)$, where the minimum one is consistent with the expression (5.20). Hence, the proof is complete.

In the main results, Theorem 5.3 provides the upper bounds of error variances in terms of two sets of Riccati-like equations. Then, Theorem 5.4 further obtains the locally optimal bound on the filtering error variance by selecting the gain parameter $K(i,j)$ at each instant. From (5.8), (5.9), (5.20), and (5.21), one concludes that all the information of measurable and uncertain probabilities of missing measurements and noise covariances has been applied to acquire the desired upper bound $M_u(i,j)$ and the filter gain $K(i,j)$, where $M_u(i,j)$ would be less conservative than those with exactly known probabilities of missing measurements.

5.3 Numerical Example

In this section, a numerical example is presented to show the effectiveness of the proposed filter design strategy.

Consider a 2-D shift-varying system (5.1) defined on a finite horizon $[0\ 40]$ with the following parameters:

$$A_1(i,j) = \begin{bmatrix} 0.4 & 0.1 + 0.1\sin(3i) \\ 0.2 & -0.35 \end{bmatrix},$$

$$A_2(i,j) = \begin{bmatrix} 0.3 + 0.1\cos(2j) & 0.1 \\ -0.2 & 0.25 \end{bmatrix},$$

$$B_1(i,j) = \begin{bmatrix} 0.15 \\ 0.1e^{-j} \end{bmatrix},$$

$$B_2(i,j) = \begin{bmatrix} 0.1 + 0.1e^{-3i} \\ 0.12 - 0.1\cos(2j) \end{bmatrix},$$

$$C(i,j) = [1.5\quad -1].$$

The noises $w(i,j)$ and $v(i,j)$ are modeled by two uncorrelated Gaussian white sequences with zero mean and variances $Q(i,j) = 0.16$ and $R(i,j) = 0.09$, respectively. The determined part of the probability for missing measurement is given as $\bar{\lambda}_1 = 0.65$. In this example, the scaling scalars are chosen as $\mu = \gamma = \epsilon = 1$, and initial conditions are set as $u_1(i) = u_2(j) = [0\ 0]^T$, $P_u(i,0) = P_u(0,j) = 0.1I$.

For simplicity, $\tilde{x}^{(l)}(i,j)$ ($l = 1,2$) denotes the l-th element of vector $\tilde{x}(i,j)$, and $M_u^{(ll)}(i,j)$ and $P_u^{(ll)}(i,j)$ stand for the (l,l)-th elements of matrices $M_u(i,j)$ and $P_u(i,j)$, respectively. According to the developed filtering strategy, the filter gains and the error variances can be recursively calculated. Part of the results are listed in Table 5.1, where the other solutions are omitted for space consideration. By applying the Monte Carlo method to 100 independent trials, the simulation results are presented in Figs. 5.1–5.6. To be specific, for $l = 1,2$, Figs. 5.1–5.2 show the estimation errors $\tilde{x}^{(l)}(i,j)$ of the considered system, and Figs. 5.5–5.6 exhibit the differences between the desired upper bounds $M_u^{(ll)}(i,j)$ and the accurate error variances $P_u^{(ll)}(i,j)$. From Figs. 5.1–5.6, we know that the proposed filter performs well with an acceptable filtering performance. We could expect the compactness of the derived upper bound according to the specific efforts devoted to the uncertain probabilities of missing measurements. We point out that the recursive filter scheme under consideration is regarded as the Kalman filter in 2-D case which, unfortunately, has been largely neglected for 2-D shift-varying systems. Actually, available filtering results on 2-D shift-varying systems have been quite scattered, especially for the variance-constrained one subject to missing measurements.

TABLE 5.1: Part of the filter gains and upper bounds

Instant	$(i,j) = (1,2)$			$(i,j) = (2,1)$			$\cdots$
$K(i,j)$		0.4069 -0.2227			0.3454 -0.1401		$\cdots$
$M_p(i,j)$	0.0611 -0.0008	-0.0008 0.0493		0.0563 0.0157	0.0157 0.0514		$\cdots$
$M_u(i,j)$	0.0732 0.0251	0.0251 0.0840		0.0817 0.0439	0.0439 0.0977		$\cdots$

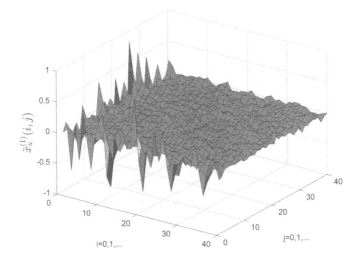

FIGURE 5.1: Estimation error $\tilde{x}_u^{(1)}(i,j)$.

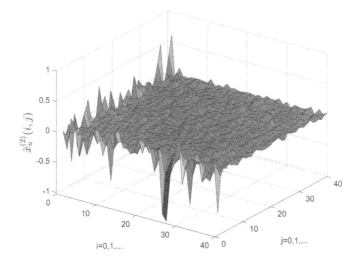

FIGURE 5.2: Estimation error $\tilde{x}_u^{(2)}(i,j)$.

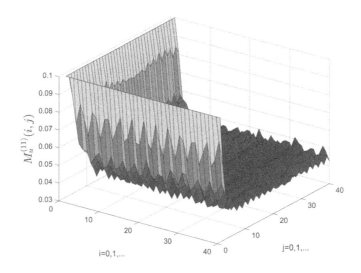

FIGURE 5.3: Trajectory of the upper bound $M_u^{(11)}(i,j)$.

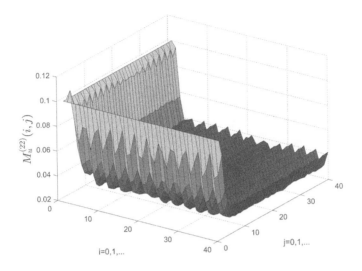

FIGURE 5.4: Trajectory of the upper bound $M_u^{(22)}(i,j)$.

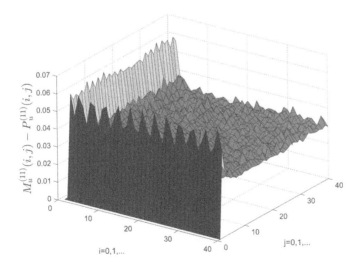

FIGURE 5.5: The difference between $M_u^{(11)}(i,j)$ and $P_u^{(11)}(i,j)$.

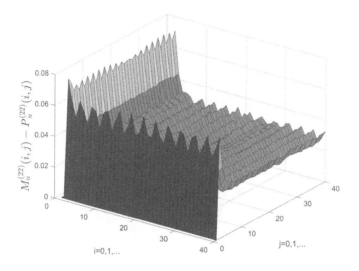

FIGURE 5.6: The difference between $M_u^{(22)}(i,j)$ and $P_u^{(22)}(i,j)$.

5.4 Summary

In this chapter, we have investigated the recursive filtering problem for 2-D shift-varying systems subject to missing measurements. The packet arrival probabilities are unknown which consist of not only the determined part but also the uncertain one. Instead of obtaining the exact filtering error variance, an upper bound has been constructed for the filtering error variance owing to the occurrence of uncertainties. Then, the developed upper bound has been minimized at each instant with the desired filter parameter. A simulation example has also been given to reveal the usefulness of the proposed recursive filter design scheme.

6

Resilient State Estimation for 2-D Shift-Varying Systems with Redundant Channels

An active research topic in relation to 2-D shift-varying systems is the state estimation issue which aims at acquiring the true states based on the available but possibly corrupted measurements. For typical state estimation problems, there have been different performance indices on the estimation errors which, in turn, have led to different estimation algorithms. Particularly, the recursive filtering algorithm focuses on the design of filter gains to obtain the minimal estimation error variance or a guaranteed upper bound at each instant. Such kind of filtering algorithm is crucial to ensure an acceptable variance-constrained performance, especially for the case where the system model undergoes incomplete information (e.g., parameter uncertainties or imperfect measurements).

In most existing literature, it has been implicitly assumed that the gain parameter of the designed estimator can be accurately realized [243, 269]. Such an assumption is, unfortunately, a bit too stringent since the actual implementation of a purposely designed filter gain may suffer from the round-off error caused by a number of issues such as the limited word length in numerical computations, the imprecision in the resolution of instrumentation and the analogue-digital conversion [55, 179]. As pointed out in [88], even small variations/drifts in gain parameters could lead to serious performance degradation. To better attenuate the influence from gain variations, the so-called resilient (or non-fragile) state estimation scheme has been proposed to ensure the insensitivity of the desired estimator against certain implementation errors. Up to now, there has been a wealth of literature on the resilience of estimators, filters or controllers designed for 1-D systems, see e.g., [76, 131, 246, 247, 257]. When it comes to the 2-D settings, the corresponding results for the resilient estimation problem with gain perturbation have been very few if not none.

In networked control systems, the packet dropouts have received an increasing research interest because of their passive influence on the system performance. An excellent way to amend the undesired effect of the packet dropouts is to enhance the network reliability by providing multiple channels for data transmissions. For this purpose, the redundant-channel transmission has been proposed [137]. To be specific, redundant channels mean that two or more channels can be accessible during the information exchange. Once the signal to be sent has no access to the first channel, the next one will be activated to reduce the packet dropout rates and thus retain the communication reliability. Despite some initial research interest on the 1-D systems with redundant channels, very little effort has been paid to the 2-D estimation problem, let alone the case that the resilient estimator is also concerned.

Motivated from the above discussions, this chapter makes one of the first few attempts to investigate the resilient state estimation problem for 2-D shift-varying systems with redundant channels. The redundant channels under consideration are subject to multiple packet dropouts modeled by uncorrelated Bernoulli distributed sequences. The resilience performance is enforced to reflect the attenuation of the negative impact from the random fluctuation of the actually implemented estimator gain. In comparison with the existing results on the 2-D estimation problem, the main contributions of this chapter are highlighted as follows: (1) the recursive state estimation is proposed, for the first time, for 2-D systems with gain variations and redundant transmissions, (2) both mathematical induction and variance-constrained method are utilized to ensure the local minimization of certain upper bound on the estimation error variance, and (3) an effective algorithm is developed, via solving a set of Riccati-like difference equations, to determine the desired estimator, which can be implemented iteratively for online computation.

The remainder of this chapter is organized as follows. Section 6.1 formulates the 2-D resilient state estimation problem. Section 6.2 provides the design scheme of the proposed estimator and the corresponding algorithm. Section 6.3 presents a simulation example to illustrate effectiveness of the established results. Conclusions are drawn in Section 6.4.

6.1   Problem Formulation and Preliminaries

Consider the following 2-D discrete shift-varying system defined on a finite horizon $i, j \in [0 \ K]$:

$$
\begin{aligned}
x(i+1, j+1) =&A_1(i+1, j)x(i+1, j) + A_2(i, j+1)x(i, j+1) \\
&+ B_1(i+1, j)w(i+1, j) + B_2(i, j+1)w(i, j+1), \quad (6.1)
\end{aligned}
$$

where K is a given positive integer, $x(i, j) \in \mathbb{R}^{n_x}$ is the state vector, $A_l(i, j)$ and $B_l(i, j)$ $(l = 1, 2)$ are known shift-varying matrices with appropriate dimensions, and $w(i, j) \in \mathbb{R}^{n_w}$ denotes the white process noise obeying the zero-mean Gaussian distribution with variance $Q(i, j) \geq 0$.

Note that the frequently encountered phenomenon of packet dropouts may seriously degrade the system performance. As an effective transmission protocol, redundant channels are utilized in this chapter to compensate possible side effects resulting from packet dropouts. In this case, the actually received observation subject to N redundant channels is modeled by

$$y(i,j) = \alpha_1(i,j)C_1(i,j)x(i,j) + v(i,j)$$
$$+ \sum_{s=2}^{N} \left\{ \prod_{r=1}^{s-1}(1 - \alpha_r(i,j))\alpha_s(i,j)C_s(i,j)x(i,j) \right\}, \quad i,j \in [0\ K], \quad (6.2)$$

where $y(i,j) \in \mathbb{R}^{n_y}$ is the measured output, and $C_s(i,j)$ ($s \in [1\ N]$) are known shift-varying matrices with appropriate dimensions, the measurement noise $v(i,j) \in \mathbb{R}^{n_y}$ is a Gaussian white sequence with zero mean and variance $R(i,j) > 0$. Random variables $\alpha_s(i,j)$ with $s \in [1\ N]$ are N mutually uncorrelated Bernoulli distributed white sequences satisfying

$$\text{Prob}\{\alpha_s(i,j) = 1\} = \bar{\alpha}_s, \quad \text{Prob}\{\alpha_s(i,j) = 0\} = 1 - \bar{\alpha}_s,$$

where $\bar{\alpha}_s$ are known constants on the interval $[0,1]$.

Remark 6.1 *Redundant channels are taken into account, for the first time, in the context of 2-D systems with hope to enhance the reliability of signal transmission, thereby improving the estimation performance. From an engineering point of view, it is reasonable to rank the priority by making a sequence of redundant channels according to their packet arrival rates. For the measurement model (6.2), in particular, it is supposed that $\{\bar{\alpha}_s\}_{s=1}^{N}$ is a non-increasing scalar sequence, where $\bar{\alpha}_s$ indicates the arrival probability of the s-th channel. That is to say, information is always transmitted through available channels with higher packet arrival probabilities.*

Remark 6.2 *Model (6.2) depicts the measurement suffering from packet dropout through N redundant channels. Such a model could cover many different cases for the actual output. Particularly, if the variable $\alpha_1(i,j)$ is set as 1, the measured output is then presented as $y(i,j) = C_1x(i,j) + v(i,j)$, which means that the first channel works normally while all the other channels are not used. Hence, the signal is observed just from the first channel. For a given integer $s \in [2\ N]$, if $\alpha_s(i,j) = 1$ and variables $\alpha_r(i,j) = 0$ for all $r \in [1\ s-1]$, then signals are transmitted through the s-th channel with channels from the first one to the $(s-1)$-th one subjecting to packet dropout. In addition, the case that $\alpha_s(i,j) = 0$ for all $s \in [1\ N]$ means the worst case where none of the channels works but only noises are received.*

For presentation convenience, we define

$$C(i,j) \triangleq [C_1^T(i,j)\ C_2^T(i,j) \cdots C_N^T(i,j)]^T, \quad \gamma_1(i,j) \triangleq \alpha_1(i,j)$$
$$\bar{\gamma}_1 \triangleq \bar{\alpha}_1, \quad \tilde{\gamma}_1(i,j) \triangleq \gamma_1(i,j) - \bar{\gamma}_1, \quad \tilde{\gamma}_s(i,j) \triangleq \gamma_s(i,j) - \bar{\gamma}_s, \quad s \in [2\ N]$$
$$\gamma_s(i,j) \triangleq \prod_{r=1}^{s-1}(1 - \alpha_r(i,j))\alpha_s(i,j), \quad \bar{\gamma}_s \triangleq \prod_{r=1}^{s-1}(1 - \bar{\alpha}_r)\bar{\alpha}_s,$$
$$\Gamma(i,j) \triangleq [\gamma_1(i,j)I_{n_y}\ \gamma_2(i,j)I_{n_y} \cdots \gamma_N(i,j)I_{n_y}],$$

$$\tilde{\Gamma}(i,j) \triangleq [\tilde{\gamma}_1(i,j)I_{n_y} \ \tilde{\gamma}_2(i,j)I_{n_y} \dots \tilde{\gamma}_N(i,j)I_{n_y}],$$

$$\bar{\Gamma} \triangleq [\gamma_1 I_{n_y} \ \gamma_2 I_{n_y} \dots \gamma_N I_{n_y}].$$

Then, the following compact system is obtained from (6.1) to (6.2):

$$x(i+1,j+1) = A_1(i+1,j)x(i+1,j) + A_2(i,j+1)x(i,j+1)$$
$$+ B_1(i+1,j)w(i+1,j) + B_2(i,j+1)w(i,j+1), \quad (6.3a)$$
$$y(i,j) = \Gamma(i,j)C(i,j)x(i,j) + v(i,j), \quad (6.3b)$$

where the initial conditions are given as $\mathbb{E}\{x(i,0)\} = u_1(i)$ and $\mathbb{E}\{x(0,j)\} = u_2(j)$ for $i,j \in [0 \ K]$, in which $u_1(i)$ and $u_2(j)$ are known vectors with $u_1(0) = u_2(0) = 0$.

A two-step estimator is proposed for (6.3) as follows:

$$\hat{x}_p(i,j) = A_1(i,j-1)\hat{x}_u(i,j-1) + A_2(i-1,j)\hat{x}_u(i-1,j), \quad (6.4a)$$
$$\hat{x}_u(i,j) = \hat{x}_p(i,j) + (K(i,j) + \Delta K(i,j)) \left[y(i,j) - \bar{\Gamma}C(i,j)\hat{x}_p(i,j) \right], \quad (6.4b)$$

where, for $i,j \in [1 \ K]$, $\hat{x}_p(i,j)$ and $\hat{x}_u(i,j)$ are the one-step prediction and the estimate of state $x(i,j)$, respectively. Matrix $K(i,j)$ is the estimator gain to be determined while $\Delta K(i,j)$ indicates the gain perturbation satisfying

$$\mathbb{E}\{\Delta K(i,j)\} = 0, \quad (6.5a)$$
$$\mathbb{E}\{\Delta K(i,j)\Delta K^T(i,j)\} \leq m(i,j)I, \quad (6.5b)$$

where $m(i,j) > 0$ is a given scalar. Initial conditions related to system (6.4) are set as $\hat{x}_u(i,0) = u_1(i)$ and $\hat{x}_u(0,j) = u_2(j)$ for $i,j \in [0 \ K]$.

Remark 6.3 *Owing to the existence of unavoidable modeling errors and digitalization-induced truncation errors, it is difficult to obtain the accurate estimator parameters that globally minimize the estimation error variance. As such, a seemingly natural way is to describe the estimator as (6.4) with gain variation $\Delta K(i,j)$. Statistical constraints on $\Delta K(i,j)$ are given in (6.5), where (6.5a) infers that the stochastic perturbation is fluctuated around the exact gain parameter $K(i,j)$, and the second-order moment of $\Delta K(i,j)$ characterizes the range of the variation for the stochastic perturbation.*

Throughout this chapter, the following assumptions are made.

Assumption 6.1 *The following statistical property holds for all $i,j,k,l \in [0 \ K]$ and $s,t \in [1 \ N]$:*

$$\mathbb{E}\left\{\eta_s(i,j)\eta_t^T(k,l)\right\}$$
$$= \text{diag}\left\{\bar{\alpha}_s(1-\bar{\alpha}_s)\delta(s,t), Q(i,j), R(i,j), X(i,0), X(0,j)\right\}\delta(i,k)\delta(j,l),$$

where $\tilde{\alpha}_s(i,j) = \alpha_s(i,j) - \bar{\alpha}_s$ and

$$\eta_s(i,j) = \left[\tilde{\alpha}_s^T(i,j) \ w^T(i,j) \ v^T(i,j) \ x^T(i,0) \ x^T(0,j)\right]^T,$$

$X(i,0)$ and $X(0,j)$ are known second-order moments of initial states, and $\delta(i,k)$ is the Kronecker delta function.

Assumption 6.2 *For $i, j \in [0\ K]$, the random matrix $\Delta K(i, j)$ is uncorrelated with all the other stochastic variables as well as the initial states.*

Define the prediction error and the estimation error as $\tilde{x}_p(i, j) \triangleq x(i, j) - \hat{x}_p(i, j)$ and $\tilde{x}_u(i, j) \triangleq x(i, j) - \hat{x}_u(i, j)$, respectively. Then, the following error dynamics is obtained from (6.3) to (6.4):

$$\tilde{x}_p(i, j) = A_1(i, j-1)\tilde{x}_u(i, j-1) + A_2(i-1, j)\tilde{x}_u(i-1, j)$$
$$+ B_1(i, j-1)w(i, j-1) + B_2(i-1, j)w(i-1, j), \tag{6.6}$$
$$\tilde{x}_u(i, j) = \left[I - (K(i, j) + \Delta K(i, j))\bar{\Gamma}C(i, j)\right]\tilde{x}_p(i, j)$$
$$- (K(i, j) + \Delta K(i, j))\left[\tilde{\Gamma}(i, j)C(i, j)x(i, j) + v(i, j)\right]. \tag{6.7}$$

The objective of this chapter is to design an estimator of the form (6.4) such that, for the considered 2-D shift-varying system (6.1), not only the existence of an upper bound is guaranteed on the estimation error variance, but also such an upper bound can be locally minimized by appropriately selecting the gain parameter at each instant.

To conclude this section, the following lemma is introduced.

Lemma 6.1 *For matrices $\tilde{\Gamma}(i, j)$ and $C(i, j)$, let $X \in \mathbb{R}^{n_x \times n_x}$ be a real-valued matrix, then the following equality*

$$\mathbb{E}\{\tilde{\Gamma}(i, j)C(i, j)XC^T(i, j)\tilde{\Gamma}^T(i, j)\} = \bar{C}(i, j)(\hat{\Gamma} \otimes X)\bar{C}^T(i, j) \tag{6.8}$$

is true, where

$$\bar{C}(i, j) = [C_1(i, j)\ C_2(i, j) \ldots C_N(i, j)], \quad \hat{\Gamma} = (\gamma_{kl})_{N \times N}$$

in which

$$\gamma_{kl} = \begin{cases} \bar{\gamma}_k(1 - \bar{\gamma}_k), & k = l \\ -\bar{\gamma}_k\bar{\gamma}_l, & k \neq l. \end{cases}$$

Proof *From the definition of random variables $\gamma_k(i, j)$ with $k \in [1\ N]$ and $i, j \in [0\ K]$, it is evident to show that*

$$\mathbb{E}\{\gamma_k(i, j)\} = \mathbb{E}\left\{\gamma_k^2(i, j)\right\} = \bar{\gamma}_k,$$
$$\mathbb{E}\{\gamma_k(i, j)\gamma_l(i, j)\} = 0, \quad k \neq l.$$

Then, based on the expression of $\tilde{\gamma}_k(i, j)$ with $k \in [1\ N]$ and $i, j \in [0\ K]$, one has

$$\mathbb{E}\left\{\tilde{\gamma}_k^2(i, j)\right\} = \bar{\gamma}_k(1 - \bar{\gamma}_k),$$
$$\mathbb{E}\left\{\tilde{\gamma}_k(i, j)\tilde{\gamma}_l(i, j)\right\} = -\bar{\gamma}_k\bar{\gamma}_l, \quad k \neq l.$$

It follows from the property of the Kronecker product that

$$\mathbb{E}\{\tilde{\Gamma}(i, j)C(i, j)XC^T(i, j)\tilde{\Gamma}^T(i, j)\}$$

$$= \mathbb{E}\Big\{\Big(\sum_{s=1}^{N}\tilde{\gamma}_s(i,j)C_s(i,j)\Big)X\Big(\sum_{t=1}^{N}\tilde{\gamma}_t(i,j)C_t(i,j)\Big)^T\Big\}$$

$$= \mathbb{E}\Big\{\sum_{s=1}^{N}\sum_{t=1}^{N}C_s(i,j)^T X C_t^T(i,j)\Big\}$$

$$= \bar{C}(i,j)(\hat{\Gamma}\otimes X)\bar{C}^T(i,j)$$

which completes the proof.

6.2   Resilient Filter Design

In this section, an upper bound will be constructed for the estimation error variance and the estimator will be designed to optimize such a bound. Before establishing the main results, the unbiasedness of the estimator is to be discussed and the dynamics is to be analyzed for the second-order moment of the system states, the one-step prediction error variance and the estimation error variance.

Define

$$X(i,j) \triangleq \mathbb{E}\left\{x(i,j)x^T(i,j)\right\},$$

$$P_p(i,j) \triangleq \mathbb{E}\left\{\tilde{x}_p(i,j)\tilde{x}_p^T(i,j)\right\},$$

$$P_u(i,j) \triangleq \mathbb{E}\left\{\tilde{x}_u(i,j)\tilde{x}_u^T(i,j)\right\}.$$

The following lemmas are established that will be utilized in our later developments.

Lemma 6.2 *The adopted estimator (6.4) is unbiased, i.e.,* $\mathbb{E}\{\tilde{x}_u(i,j)\} = 0$ *for* $i,j \in [0\ K]$ *and* $\mathbb{E}\{\tilde{x}_p(i,j)\} = 0$ *for* $i,j \in [1\ K]$.

Proof *The conclusion is drawn based on the inductive approach. According to the initial conditions, the validity of the following relationship*

$$\mathbb{E}\left\{\tilde{x}_u(i,0)\right\} = \mathbb{E}\left\{\tilde{x}_u(0,j)\right\} = 0 \tag{6.9}$$

holds for $i,j \in [0\ K]$. *Then, it follows from (6.6) and (6.9) that*

$$\mathbb{E}\left\{\tilde{x}_p(1,1)\right\} = A_1(1,0)\mathbb{E}\left\{\tilde{x}_u(1,0)\right\} + A_2(0,1)\mathbb{E}\left\{\tilde{x}_u(0,1)\right\} = 0.$$

Recalling the statistics presented in (6.5) and Assumptions 6.1–6.2, we derive the following equality from (6.7):

$$\mathbb{E}\left\{\tilde{x}_u(1,1)\right\} = \left[I - K(1,1)\bar{\Gamma}C(1,1)\right]\mathbb{E}\left\{\tilde{x}_p(1,1)\right\} = 0$$

which implies that the relationship

$$\mathbb{E}\left\{\tilde{x}_p(i,j)\right\} = \mathbb{E}\left\{\tilde{x}_u(i,j)\right\} = 0$$

is true for $(i,j) \in \{(i_0, j_0)|i_0, j_0 \geq 1;\ i_0 + j_0 = 2\}$.

To proceed further, we suppose that $\mathbb{E}\{\tilde{x}_u(i,j)\} = 0$ *holds for* $(i,j) \in \{(i_0, j_0)|i_0, j_0 \in [0\ k];\ i_0 + j_0 = k\}$, *where* $k \in [1\ 2K-1]$ *is a given integer. Then, for* $(i,j) \in \{i_0, j_0 \geq 1;\ i_0 + j_0 = k+1\}$ *with the given integer* $k \in [1\ 2K-1]$, *the inductive hypothesis, together with (6.6), results in*

$$\mathbb{E}\left\{\tilde{x}_p(i,j)\right\} = A_1(i,j-1)\mathbb{E}\left\{\tilde{x}_u(i,j-1)\right\} + A_2(i-1,j)\mathbb{E}\left\{\tilde{x}_u(i-1,j)\right\} = 0$$

which further means

$$\mathbb{E}\left\{\tilde{x}_u(i,j)\right\} = \left[I - K(i,j)\bar{\Gamma}C(i,j)\right]\mathbb{E}\left\{\tilde{x}_p(i,j)\right\} = 0.$$

The conclusion follows immediately due to the mathematical induction.

Lemma 6.3 *For system (6.3), the recursion of the second-order moment of the state with* $i,j \in [1\ K]$ *is given as follows:*

$$\begin{aligned}
X(i,j) =\ & A_1(i,j-1)X(i,j-1)A_1^T(i,j-1) \\
& + A_1(i,j-1)\mathbb{E}\{x(i,j-1)x^T(i-1,j)\}A_2^T(i-1,j) \\
& + A_2(i-1,j)X(i-1,j)A_2^T(i-1,j) \\
& + A_2(i-1,j)\mathbb{E}\{x(i-1,j)x^T(i,j-1)\}A_1^T(i,j-1) \\
& + B_1(i,j-1)Q(i,j-1)B_1^T(i,j-1) \\
& + B_2(i-1,j)Q(i-1,j)B_2^T(i-1,j).
\end{aligned} \tag{6.10}$$

Proof *Owing to the statistical property of random variables* $w(i,j)$, *(6.10) is obtained readily from (6.3). The details are omitted here for brevity.*

Remark 6.4 *Note that the cross term* $\mathbb{E}\{x(i-1,j)x^T(i,j-1)\}$ *is involved in (6.10) whose evolution needs further analysis. For* $i,j \in [2\ K]$, *it follows from (6.3) that*

$$\begin{aligned}
& \mathbb{E}\left\{x(i,j-1)x^T(i-1,j)\right\} \\
& = A_1(i,j-2)\mathbb{E}\left\{x(i,j-2)x^T(i-1,j-1)\right\}A_1^T(i-1,j-1) \\
& \quad + A_1(i,j-2)\mathbb{E}\left\{x(i,j-2)x^T(i-2,j)\right\}A_2^T(i-2,j) \\
& \quad + A_2(i-1,j-1)\mathbb{E}\left\{x(i-1,j-1)x^T(i-2,j)\right\}A_2^T(i-2,j) \\
& \quad + A_2(i-1,j-1)X(i-1,j-1)A_1^T(i-1,j-1) \\
& \quad + B_2(i-1,j-1)Q(i-1,j-1)B_1^T(i-1,j-1).
\end{aligned} \tag{6.11}$$

In addition, for $k \in [2\ K-1]$ *and* $i,j \in [k+1\ K]$, *one has*

$$\mathbb{E}\left\{x(i,j-k)x^T(i-k,j)\right\}$$

$$
\begin{aligned}
= \ &A_1(i,j-k-1)\mathbb{E}\left\{x(i,j-k-1)x^T(i-k,j-1)\right\}A_1^T(i-k,j-1)\\
&+ A_1(i,j-k-1)\mathbb{E}\left\{x(i,j-k-1)x^T(i-k-1,j)\right\}A_2^T(i-k-1,j)\\
&+ A_2(i-1,j-k)\mathbb{E}\left\{x(i-1,j-k)x^T(i-k,j-1)\right\}A_1^T(i-k,j-1)\\
&+ A_2(i-1,j-k)\mathbb{E}\left\{x(i-1,j-k)x^T(i-k-1,j)\right\}A_2^T(i-k-1,j).
\end{aligned}
$$
(6.12)

Together with the given statistics of initial states, equations (6.10)–(6.12) provide a complete recursion suitable for calculating $X(i,j)$ recursively.

Lemma 6.4 *The evolutions of the prediction error variance $P_p(i,j)$ and the estimation error variance $P_u(i,j)$ satisfy, respectively, the following Riccati-like equations:*

$$
\begin{aligned}
P_p(i,j) =\ &A_1(i,j-1)P_u(i,j-1)A_1^T(i,j-1)\\
&+ A_2(i-1,j)P_u(i-1,j)A_2^T(i-1,j)\\
&+ A_1(i,j-1)\mathbb{E}\left\{\tilde{x}_u(i,j-1)\tilde{x}_u^T(i-1,j)\right\}A_2^T(i-1,j)\\
&+ A_2(i-1,j)\mathbb{E}\left\{\tilde{x}_u(i-1,j)\tilde{x}_u^T(i,j-1)\right\}A_1^T(i,j-1)\\
&+ B_1(i,j-1)Q(i,j-1)B_1^T(i,j-1)\\
&+ B_2(i-1,j)Q(i-1,j)B_2^T(i-1,j),
\end{aligned}
$$
(6.13)

$$
\begin{aligned}
P_u(i,j) =\ &\left[I-K(i,j)\bar{\Gamma}C(i,j)\right]P_p(i,j)\left[I-K(i,j)\bar{\Gamma}C(i,j)\right]^T\\
&+ K(i,j)\bar{R}(i,j)K^T(i,j)+\mathbb{E}\left\{\Delta K(i,j)\bar{\Pi}(i,j)\Delta K^T(i,j)\right\}
\end{aligned}
$$
(6.14)

for $i,j\in[1\ K]$, where

$$
\bar{R}(i,j)=\bar{C}(i,j)(\hat{\Gamma}\otimes X(i,j))\bar{C}^T(i,j)+R(i,j),
$$
$$
\bar{\Pi}(i,j)=\bar{\Gamma}C(i,j)P_p(i,j)C^T(i,j)\bar{\Gamma}^T+\bar{R}(i,j).
$$

Proof *Combining (6.6) with the definition of $P_p(i,j)$, one can confirm the validity of (6.13) easily. Moreover, recall the fact that $\tilde{x}_p(i,j)$, $\Delta K(i,j)$, $\tilde{\Gamma}(i,j)$, and $v(i,j)$ are mutually uncorrelated with each other, as well as the statistic properties $\mathbb{E}\{\Delta K(i,j)\}=0$, $\mathbb{E}\{\tilde{\Gamma}(i,j)\}=0$, and $\mathbb{E}\{v(i,j)\}=0$. One obtains from (6.7), the expression of $P_u(i,j)$, and Lemma 6.1 that*

$$
\begin{aligned}
P_u(i,j) =\ &\mathbb{E}\Big\{\left[I-(K(i,j)+\Delta K(i,j))\bar{\Gamma}C(i,j)\right]P_p(i,j)\\
&\times\left[I-(K(i,j)+\Delta K(i,j))\bar{\Gamma}C(i,j)\right]^T\Big\}\\
&+\mathbb{E}\left\{(K(i,j)+\Delta K(i,j))\bar{R}(i,j)(K(i,j)+\Delta K(i,j))^T\right\}\\
=\ &\left[I-K(i,j)\bar{\Gamma}C(i,j)\right]P_p(i,j)\left[I-K(i,j)\bar{\Gamma}C(i,j)\right]^T\\
&+ K(i,j)\bar{R}(i,j)K^T(i,j)+\mathbb{E}\left\{\Delta K(i,j)\bar{\Pi}(i,j)\Delta K^T(i,j)\right\}.
\end{aligned}
$$

Consequently, the recursion of $P_u(i,j)$ is obtained as shown in (6.14), which ends the proof.

The above lemma presents the recursions of the prediction and the estimation error variances. As the gain variation $\Delta K(i,j)$ is involved in (6.14), it is impossible to calculate the explicit solution of the estimation error variance $P_u(i,j)$. Therefore, an alternative yet effective way is proposed below to find out an upper bound for $P_u(i,j)$ and then achieve the minimal one at each step by designing an appropriate gain parameter.

Theorem 6.1 *Let $\mu > 0$ be a given positive scalar. Assume that there are two sequences of positive matrices $\{S_p(i,j)\}$ and $\{S_u(i,j)\}$ such that the following two recursive equations*

$$
\begin{aligned}
S_p(i,j) = &(1+\mu)A_1(i,j-1)S_u(i,j-1)A_1^T(i,j-1) \\
&+ (1+\mu^{-1})A_2(i-1,j)S_u(i-1,j)A_2^T(i-1,j) \\
&+ B_1(i,j-1)Q(i,j-1)B_1^T(i,j-1) \\
&+ B_2(i-1,j)Q(i-1,j)B_2^T(i-1,j),
\end{aligned}
\tag{6.15}
$$

$$
\begin{aligned}
S_u(i,j) = &\left[I - K(i,j)\bar{\Gamma}C(i,j)\right]S_p(i,j)\left[I - K(i,j)\bar{\Gamma}C(i,j)\right]^T \\
&+ K(i,j)\bar{R}(i,j)K^T(i,j) + \lambda_{\max}\{\tilde{\Pi}(i,j)\}m(i,j)I
\end{aligned}
\tag{6.16}
$$

with initial constraints

$$
S_u(i,0) = P_u(i,0), \quad S_u(0,j) = P_u(0,j), \quad i,j \in [1\ K]
$$

are satisfied, where

$$
\tilde{\Pi}(i,j) = \bar{\Gamma}C(i,j)S_p(i,j)C^T(i,j)\bar{\Gamma}^T + \bar{R}(i,j).
\tag{6.17}
$$

Then, matrices $P_p(i,j)$ and $P_u(i,j)$ in (6.13)-(6.14) are bounded by

$$
P_p(i,j) \le S_p(i,j), \quad P_u(i,j) \le S_u(i,j), \quad i,j \in [1\ K].
$$

Proof *For $i,j \in [1\ K]$ and any positive scalar $\mu > 0$, it follows from (6.13) that*

$$
\begin{aligned}
P_p(i,j) \le &(1+\mu)A_1(i,j-1)P_u(i,j-1)A_1^T(i,j-1) \\
&+ (1+\mu^{-1})A_2(i-1,j)P_u(i-1,j)A_2^T(i-1,j) \\
&+ B_1(i,j-1)Q(i,j-1)B_1^T(i,j-1) \\
&+ B_2(i-1,j)Q(i-1,j)B_2^T(i-1,j).
\end{aligned}
\tag{6.18}
$$

Moreover, (6.14) together with (6.5) gives rise to the following inequality for $i,j \in [1\ K]$:

$$
\begin{aligned}
P_u(i,j) \le &\left[I - K(i,j)\bar{\Gamma}C(i,j)\right]P_p(i,j)\left[I - K(i,j)\bar{\Gamma}C(i,j)\right]^T \\
&+ K(i,j)\bar{R}(i,j)K^T(i,j) + \lambda_{\max}\{\tilde{\Pi}(i,j)\}m(i,j)I.
\end{aligned}
\tag{6.19}
$$

In what follows, this theorem will be confirmed by the induction method. From (6.15), (6.18), and the initial conditions, one has

$$P_p(1,1) - S_p(1,1)$$
$$\leq (1+\mu)A_1(1,0)[P_u(1,0) - S_u(1,0)]A_1^T(1,0)$$
$$+ (1+\mu^{-1})A_2(0,1)[P_u(0,1) - S_u(0,1)]A_2^T(0,1)$$
$$\leq 0 \tag{6.20}$$

which yields

$$P_u(1,1) - S_u(1,1)$$
$$\leq \left[I - K(1,1)\bar{\Gamma}C(1,1)\right][P_p(1,1) - S_p(1,1)]\left[I - K(1,1)\bar{\Gamma}C(1,1)\right]^T$$
$$+ \left[\lambda_{\max}\{\bar{\Pi}(1,1)\} - \lambda_{\max}\{\tilde{\Pi}(1,1)\}\right]m(1,1)I$$
$$\leq 0. \tag{6.21}$$

That is to say, the validity of $P_u(i,j) \leq S_u(i,j)$ holds for $(i,j) \in \{(i_0,j_0)|i_0,j_0 \geq 1; \; i_0 + j_0 = 2\}$.

Assume that $P_u(i,j) \leq S_u(i,j)$ is satisfied for all $(i,j) \in \{(i_0,j_0)|i_0,j_0 \in [1 \;\; k-1]; \; i_0 + j_0 = k\}$ with a given integer $k \in [2 \;\; 2K-1]$. Then, for $(i,j) \in \{(i_0,j_0)|i_0,j_0 \in [1 \;\; k]; \; i_0 + j_0 = k+1\}$, we arrive at

$$P_p(i,j) - S_p(i,j)$$
$$\leq (1+\mu)A_1(i,j-1)[P_u(i,j-1) - S_u(i,j-1)]A_1^T(i,j-1)$$
$$+ (1+\mu^{-1})A_2(i-1,j)[P_u(i-1,j) - S_u(i-1,j)]A_2^T(i-1,j)$$
$$\leq 0 \tag{6.22}$$

which further indicates

$$P_u(i,j) - S_u(i,j)$$
$$\leq \left[I - K(i,j)\bar{\Gamma}C(i,j)\right][P_p(i,j) - S_p(i,j)]\left[I - K(i,j)\bar{\Gamma}C(i,j)\right]^T$$
$$+ \left[\lambda_{\max}\{\bar{\Pi}(i,j)\} - \lambda_{\max}\{\tilde{\Pi}(i,j)\}\right]m(i,j)I$$
$$\leq 0. \tag{6.23}$$

In virtue of the inductive method, the proof is complete.

Theorem 6.1 provides a sufficient condition that guarantees the existence of an upper bound on the estimation error variance, where such a bound can be derived by solving two sets of Riccati-like equations (6.15) and (6.16). The estimator (6.4) with gain parameters ensuring the minimal upper bound is to be designed in the following theorem.

Theorem 6.2 *Consider system (6.3) with a resilient state estimator (6.4).*

The upper bound $S_u(i,j)$ for the estimation error variance $P_u(i,j)$ is locally minimized with the gain parameter designed as

$$K(i,j) = S_p(i,j)C^T(i,j)\bar{\Gamma}^T\tilde{\Pi}^{-1}(i,j), \qquad (6.24)$$

where $\tilde{\Pi}(i,j)$ is defined in (6.17). In this case, the minimum of $S_u(i,j)$ is given as

$$\begin{aligned} S_u(i,j) =& S_p(i,j) + \lambda_{\max}\{\tilde{\Pi}(i,j)\}m(i,j)I \\ & - S_p(i,j)C^T(i,j)\bar{\Gamma}^T\tilde{\Pi}^{-1}(i,j)\bar{\Gamma}C(i,j)S_p^T(i,j). \end{aligned} \qquad (6.25)$$

Proof *In view of (6.16), one has*

$$\begin{aligned} S_u(i,j) =& S_p(i,j) + \lambda_{\max}\{\tilde{\Pi}(i,j)\}m(i,j)I \\ & - K(i,j)\bar{\Gamma}C(i,j)S_p(i,j) - S_p(i,j)C^T(i,j)\bar{\Gamma}^T K^T(i,j) \\ & + K(i,j)\left[\bar{\Gamma}C(i,j)S_p(i,j)C^T(i,j)\bar{\Gamma}^T + \bar{R}(i,j)\right]K^T(i,j) \\ =& S_p(i,j) + \lambda_{\max}\{\tilde{\Pi}(i,j)\}m(i,j)I \\ & - S_p(i,j)C^T(i,j)\bar{\Gamma}^T\tilde{\Pi}^{-1}(i,j)\bar{\Gamma}C(i,j)S_p^T(i,j) \\ & + \left[K(i,j) - S_p(i,j)C^T(i,j)\bar{\Gamma}^T\tilde{\Pi}^{-1}(i,j)\right]\tilde{\Pi}(i,j) \\ & \times \left[K(i,j) - S_p(i,j)C^T(i,j)\bar{\Gamma}^T\tilde{\Pi}^{-1}(i,j)\right]^T. \end{aligned} \qquad (6.26)$$

According to the completing-the-square method, the upper bound $S_u(i,j)$ reaches its minimum if and only if $K(i,j)$ is selected as (6.24), which ends the proof.

Based on the above established results, the estimator gain parameters can be calculated by solving recursions (6.10)–(6.12), (6.15), (6.24), and (6.25). The resilient estimator design algorithm is presented in Algorithm 3.

Remark 6.5 *An overall algorithm for designing the estimator is implemented based on the recursions (4), (10)–(12), (15), (24), and (25) where the available measurements are sequentially used at each step of the estimate updating, and thus the proposed Algorithm 3 facilitates for an online application. A critical concern of our estimation strategy is the computational burden of Algorithm 3. This can be calculated by the arithmetic operations in deriving the state prediction $\hat{x}_p(i,j)$, the estimate $\hat{x}_u(i,j)$, matrices $X(i,j)$, $S_p(i,j)$, $S_u(i,j)$ and the gain parameter $K(i,j)$. According to the matrix computations, at each recursion, the computational burden of Algorithm 3 is $O(n_x^3 + n_x^2 n_w + n_x n_w^2 + n_y^3 N + n_x n_y^2 N + n_y n_x^2 N^2)$, and hence the whole computational burden over the finite horizon $[0\ K] \times [0\ K]$ is given as $O((n_x^3 + n_x^2 n_w + n_x n_w^2 + n_y^3 N + n_x n_y^2 N + n_y n_x^2 N^2)K^2)$.*

Remark 6.6 *With aid of a combination of intensive stochastic analysis, mathematical induction and variance-constrained method, the resilient esti- mator is designed for the considered system such that the upper bound of the*

Algorithm 3 Resilient State Estimate Algorithm

Initialization: Give positive scalars μ and $m(i,j)$, initial conditions $u_1(i)$, $u_2(j)$, $X(i,0)$, and $X(0,j)$ for all $i,j \in [0\ K]$.

1: Set $k=1$ and $l=1$
2: **while** $k \in [1\ K]$ and $l \in [1\ K]$ **do**
3:     Calculate firstly for parameters $\hat{x}_p(k,l)$, $X(k,l)$, and $S_p(k,l)$ from (6.4a), (6.10), and (6.15), respectively; obtain secondly for matrices $K(k,l)$, $S_u(k,l)$, and estimate $\hat{x}_u(k,l)$ from (6.24), (6.25), and (6.4b), respectively.
4:     **if** $k \in [1\ K]$ and $l \in [1\ K-1]$ **then**
5:         Solve equalities (6.11) and (6.12) to obtain the cross-terms $\mathbb{E}\{x(k,l)x^T(i_0,k+l-i_0)\}$ for $i_0 \in [k+l-\min\{k+l,K\}\ \ k-1]$.
6:     **end if**
7:     **if** $k \in [1\ K]$ and $l = K$ **then**
8:         Set $k = k+1$ and $l = 1$.
9:     **else**
10:        Set $l = l+1$.
11:    **end if**
12: **end while**

---

estimation error variance is minimized at each instant. Noting that scalars $\bar{\gamma}_s$ ($s \in [1\ N]$) are the probabilities of signals successfully transmitted through the N redundant channels, variations of $\bar{\gamma}_s$ will unavoidably influence the estimation performance due to the existence of matrix $\bar{\Gamma}$ in (6.15)–(6.16). It is easy to see that the performance would be improved with higher values of $\bar{\gamma}_s$ ($s \in [1\ N]$), which signifies that more information can be observed in the sense of mean square. Otherwise, the performance is probably degraded with lower values of $\bar{\gamma}_s$.

Remark 6.7 *Randomly occurring packet dropouts may be encountered because of bandwidth constraint on transmission paths. To relieve the undesired effect, redundant channels are introduced in this chapter. Actually, the arrival probability of missing measurements can be calculated from (6.2) as:*

$$\text{Prob}\{D_N(i,j)\} = 1 - \prod_{s=1}^{N} (1 - \mathbb{E}\{\alpha_s(i,j)\}),$$

where $D_N(i,j)$ stands for the event that information is successfully transmitted through the N redundant channels at instant (i,j). For any given positive integer N, it is evident to see that

$$\text{Prob}\{D_N(i,j)\} = 1 - \prod_{s=1}^{N} (1 - \bar{\alpha}_s)$$

$$= 1 - \prod_{s=1}^{N-1} (1 - \bar{\alpha}_s) + \prod_{s=1}^{N-1} (1 - \bar{\alpha}_s) \bar{\alpha}_N$$

$$= \text{Prob} \{ D_{N-1}(i,j) \} + \prod_{s=1}^{N-1} (1 - \bar{\alpha}_s) \bar{\alpha}_N$$

$$\geq \text{Prob} \{ D_{N-1}(i,j) \}.$$

As a consequence, the communication reliability is likely to enhance as the number of redundant channels increases. Nevertheless, more redundant channels will inevitably result in a higher cost and larger power consumption. The trade-off between the reliability and resource is an interesting issue which deserves further investigation.

6.3 Numerical Examples

To illustrate the applicability of the proposed estimation strategy, an industrial thermal process is considered based on the following heat differential equation [83]:

$$\frac{\partial u(z,t)}{\partial t} + \frac{\partial u(z,t)}{\partial z} = a u(z,t) + b f(z,t),$$

where $u(z,t)$ is the temperature needed to be estimated over the space $z \in [0, Z]$ and time $t \in [0, T]$, $f(z,t)$ is a given force function, a and b are real scalars representing the heat transfer coefficients. Set $u(z,t) \triangleq u_d(i\Delta z, j\Delta t)$ for $z \in [i\Delta z, (i+1)\Delta z)$ and $t \in [j\Delta t, (j+1)\Delta t)$. The following approximate equalities are valid when properly selecting the step sizes Δz and Δt [83]:

$$\frac{\partial u(z,t)}{\partial t} \doteq \frac{u_d(i\Delta z, (j+1)\Delta t) - u_d(i\Delta z, j\Delta t)}{\Delta t},$$

$$\frac{\partial u(z,t)}{\partial z} \doteq \frac{u_d(i\Delta z, j\Delta t) - u_d((i-1)\Delta z, j\Delta t)}{\Delta z}.$$

Denote

$$\bar{u}(i,j) \triangleq u_d(i\Delta z, j\Delta t), \quad x(i,j) \triangleq [\bar{u}^T(i-1,j) \ \ \bar{u}^T(i,j)]^T.$$

Then, when $b = 0$, the original PDE can be approximated by a 2-D discrete-time FM-II model with

$$A_1 = \begin{bmatrix} 0 & 0 \\ \frac{\Delta t}{\Delta z} & 1 - \frac{\Delta t}{\Delta z} + a\Delta t \end{bmatrix}, \quad A_2 = \begin{bmatrix} 0 & 1 \\ 0 & 0 \end{bmatrix}.$$

From the practical viewpoint, the system may be contaminated by noises resulting from the unideal chemical reactor. Additionally, the considered heating model might be affected from heterogeneous media due to unpredictable changes of environments, and hence the system parameters could be shift-varying with purpose of reflecting parameter variations from time to time. Moreover, the redundant protocol is adopted to enhance the communication reliability and the possibly encountered estimator gain perturbation is also taken into account. In the sequel, set $\Delta t = 0.1$, $\Delta z = 0.4$, and $K = 40$. The shift-varying system parameters are given as $a(i, j) = \cos(0.8i) - 4$ and

$$A_1(i, j) = \begin{bmatrix} 0 & 0 \\ 0.25 & 0.35 + 0.1\cos(0.8i) \end{bmatrix},$$

$$A_2(i, j) = A_2, \quad B_1(i, j) = \begin{bmatrix} 0.2 - 0.1e^{-3i} \\ 0.15e^{-j} \end{bmatrix},$$

$$B_2(i, j) = \begin{bmatrix} 0.1 \\ 0.16 + 0.1e^{-2j} \end{bmatrix}.$$

Random variables $w(i, j)$ and $v(i, j)$ are uncorrelated zero-mean Gaussian white-noise sequences with variances $Q(i, j) = 0.49$ and $R(i, j) = 0.36$, respectively. For simulation purpose, we choose the scaling scalars as $\mu = 1$ and $m(i, j) = 0.1$, the initial conditions as $u_1(i) = u_2(j) = [0 \ 0]^T$ and $X(i, 0) = X(0, j) = 0.1I$.

We first show the effectiveness of the designed estimator strategy. Consider a single transmission channel with $C_1(i, j) = [0.35 \ 0.45 - 0.1\sin(5j)]$ and select the packet arrival probability as $\bar{\alpha}_1 = 0.9$. According to Algorithm 3, one can recursively calculate the second-order moment of state $X(i, j)$, the upper bounds $S_p(i, j)$ and $S_u(i, j)$, and the filter gain $K(i, j)$. The corresponding matrices are listed in Table 6.1. Moreover, Figs. 6.1–6.3 are given to show respectively the trajectories of the first element $\tilde{x}_u^{(1)}(i, j)$ of the estimation error $\tilde{x}_u(i, j)$, the first diagonal element $S_u^{(11)}(i, j)$ of the upper bound $S_u(i, j)$, and the difference between $S_u^{(11)}(i, j)$ and $P_u^{(11)}(i, j)$ (only part of the simulation results are presented here for brevity). It can be seen from Figs. 6.1–6.3 that the proposed scheme performs well with an acceptable estimation performance.

To qualitatively evaluate the influences of packet dropouts and redundant channels, two different scenarios are considered as follows:

(i) *The effect of the packet arrival rate on the estimation performance:* Reset the probability as $\bar{\alpha}_1 = 0.3$ without modifying any other parameters. The simulation result is plotted in Fig. 6.4. In comparison with the upper bound $S_u^{(11)}(i, j)$ with $\bar{\alpha}_1 = 0.9$, the counterpart is bigger than that with $\bar{\alpha}_1 = 0.3$, which means that the estimation performance is degraded as $\bar{\alpha}_1$ decreases. This simulation result confirms with the practical situations since a smaller value of $\bar{\alpha}_1$ represents a lower probability of packet arrival, and thus may result in less information available for obtaining the upper bound $S_u(i, j)$.

TABLE 6.1: Related matrices and gain parameters

$X(1,2) = \begin{matrix} 0.1235 & 0.0132 \\ 0.0132 & 0.0356 \end{matrix}$		$X(2,1) = \begin{matrix} 0.0741 & 0.0232 \\ 0.0232 & 0.0441 \end{matrix}$		$\cdots$
$S_p(1,2) = \begin{matrix} 0.2235 & 0.0132 \\ 0.0132 & 0.0901 \end{matrix}$		$S_p(2,1) = \begin{matrix} 0.2444 & 0.0232 \\ 0.0232 & 0.0624 \end{matrix}$		$\cdots$
$K(1,2) = \begin{matrix} 0.1877 \\ 0.1107 \end{matrix}$		$K(2,1) = \begin{matrix} 0.2159 \\ 0.0927 \end{matrix}$		$\cdots$
$S_u(1,2) = \begin{matrix} 0.2483 & 0.0142 \\ 0.0142 & 0.1100 \end{matrix}$		$S_u(2,1) = \begin{matrix} 0.2663 & 0.0150 \\ 0.0150 & 0.0998 \end{matrix}$		$\cdots$

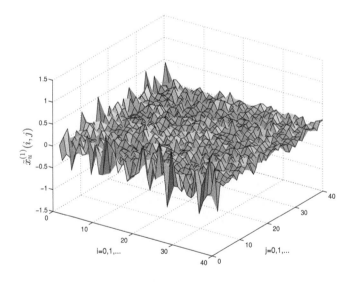

FIGURE 6.1: Estimation error $\tilde{x}_u^{(1)}(i,j)$.

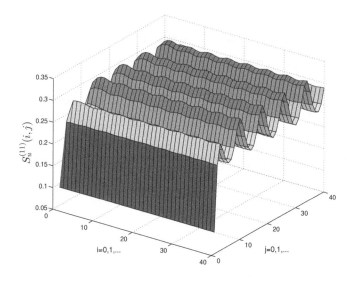

FIGURE 6.2: Evolution of the upper bound $S_u^{(11)}(i,j)$ with $\bar{\alpha}_1 = 0.9$.

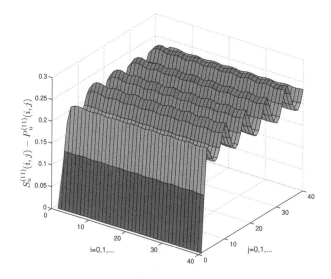

FIGURE 6.3: Difference between $S_u^{(11)}(i,j)$ and $P_u^{(11)}(i,j)$ with $\bar{\alpha}_1 = 0.9$.

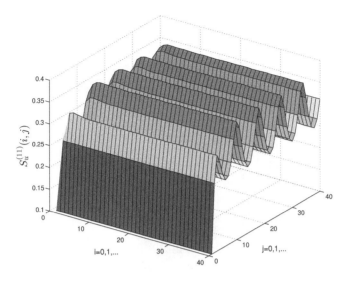

FIGURE 6.4: Evolution of the upper bound $S_u^{(11)}(i,j)$ with $\bar{\alpha}_1 = 0.3$.

(ii) *The effect of redundant channels on the estimation performance*:
In this case, we consider the 2-D system (6.1) with two redundant
channels, i.e.,

$N = 2$. The other system parameters are the same as above except that the
output matrices become

$$C_1(i,j) = C_2(i,j) = [0.35 \quad 0.45 - 0.1\sin(5j)]$$

and the packet arrival probabilities are set as $\bar{\alpha}_1 = \bar{\alpha}_2 = 0.9$. Denote $S_{u,N}^{(11)}(i,j)$
as the first diagonal element of $S_u(i,j)$ with N redundant channels. Based on
the Monte Carlo approach, Fig. 6.5 exhibits the difference between $S_{u,1}^{(11)}(i,j)$
and $S_{u,2}^{(11)}(i,j)$ obtained from 100 independent experiments. It is observed that
the estimation performance with two redundant channels is better than that
with only one channel.

   To further highlight the advantages of our estimation design scheme, the
mean square error (MSE) of the resilient estimator established in our chapter
is compared with that proposed in [269]. Define the mean square estimation
error as

$$MSE_{ij} \triangleq \frac{1}{100} \sum_{t=1}^{100} (x(i,j) - \hat{x}_u(i,j))^T (x(i,j) - \hat{x}_u(i,j))$$

and the simulation result is presented in Fig. 6.6, where MSE_{ij}^o represents the
mean square estimation error obtained from the established method in [269].
Figure 6.6 infers that our resilient estimator performs better on account of
the great efforts made on the analysis of the gain perturbations and packet
dropouts.

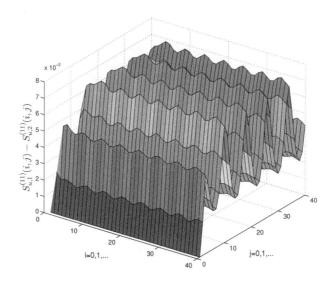

FIGURE 6.5: Difference of $S_u^{(11)}(i,j)$ with different redundant channels.

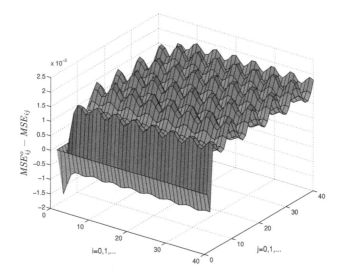

FIGURE 6.6: MSE comparison between our resilient estimator and that in [269].

6.4 Summary

In this chapter, the resilient state estimation problem has been investigated for 2-D shift-varying systems with redundant channels which are subject to randomly occurring packet dropout. A resilient estimator has been adopted with gain perturbation characterized by zero mean value and bounded second-order moment. Based on the variance-constrained method, the desired estimator has been determined under which an upper bound on the estimation error variance has been first guaranteed and then minimized. Moreover, an easy-to-implement algorithm has been established to facilitate the calculation of the desired parameters. A simulation example has also been provided to illustrate the effectiveness of the designed estimation scheme.

7

Recursive Distributed Filtering for 2-D Shift-Varying Systems Over Sensor Networks Under Random Access Protocols

---

Living in a networked world, we have been witnessing the dramatic growth of heterogeneous data produced from sensor networks in a variety of formats including text, images, videos, audio, graphics, and time series sequences. Accordingly, the data analysis problem over sensor networks has recently become an intriguing topic of research with extensive applications in practical areas such as intelligent transportation, environment monitoring, process automation, and distributed robotics [6, 21, 66, 69, 147, 245]. Generally, a sensor network consists of an array of sensing devices that are spatially distributed in certain regions. The networked sensors are intelligent nodes that are capable of wireless communication, sensing, and computation, and they communicate with neighboring nodes to perform specific tasks in a cooperative manner. A central research topic associated with sensor networks is the distributed filtering problem whose main idea is to estimate the system state by using both the local and the neighboring information under the sensor network with a certain interaction topology [22].

With a rapid revolution on network technologies, the traditional point-to-point connection has been largely replaced by the network connection through a shared network. Despite its low cost and easy implementation, the network connection may lead to data congestion problem derived from the limited network bandwidth. To circumvent such a problem, several communication protocols are deployed to regulate the data transmission order of sensor nodes. One of the widely used communication protocols is the RA protocol that is a stochastic communication strategy to identify the selection of signals/packets (involved in each sensor) entering into the network media. Inspired by some preliminary results on the 1-D filtering issues with RA protocol, an extension of the protocol-based filtering strategy to the 2-D case would be interesting.

Based on the discussions made so far, we are motivated to launch a major study on the recursive distributed filtering problem for 2-D shift-varying systems with RA scheduling. Stochastic nonlinearities possessing certain statistical properties are also considered in both the system states and the measurement outputs to reflect the ubiquitous nonlinear disturbances. The primary objective of the filtering problem concentrates on designing the distributed filters under RA scheduling to estimate the system state within a satisfactory

127

filtering performance. Such a task appears to be fairly challenging on account of the following essential difficulties: (1) how to analyze the dynamics evolution in bidirectional directions complicated by the RA scheduling and the shift-varying nature? (2) how to guarantee an upper bound on the filtering error variance as the accurate one is almost impossible to obtain due to the introduction of the communication protocol and the stochastic nonlinearities? and (3) how to deal with the network coupling issues and then design the distributed filters that locally minimize certain upper bound of the filtering error variance? We endeavor to answer these three questions in the course of the research carried out.

In this chapter, the general FM-II model is introduced to describe a class of 2-D systems with shift-varying parameters, stochastic nonlinearities, and communication protocol. The novelties of this chapter can be highlighted as follows:

(1)   A compelling RA protocol scheme will be firstly arranged for 2-D systems suffering from a capacity-constrained sensor network. The proposed RA protocol, depicted by a sequence of random variables with known probability distributions, will be utilized to determine which packet received by each sensor is allowed to be transmitted through the network at each shift step.

(2)   A refined distributed filter will be constructed for each sensor by taking the scheduling protocol and the communication topology into account. In contrast to the traditional filter design matter subject to one single sensor, the local filter designed here estimates the system state based on the aggregated innovations from not only itself but also the neighboring sensors.

(3)   Great efforts are devoted to developing an upper bound for the filtering error variance in terms of Riccati-like difference equations with the help of the inductive approach. In addition, a matrix simplification method is applied to address the challenges generated from the possible sparsity of the sensor network, thereby favorably obtaining the distributed filtering scheme in a recursive manner.

The outline of this chapter is organized as follows. Section 7.1 formulates the distributed filtering problem to be addressed. Section 7.2 presents the main results, where an upper bound of the filtering error variance is developed and then the desired filter is determined which minimizes the obtained upper bound at each iteration. A simulation example is provided in Section 7.3 to show validity of the proposed filtering algorithm. Finally, conclusions are drawn in Section 7.4.

7.1 Problem Formulation and Preliminaries

7.1.1 The System Model

Consider a class of 2-D discrete shift-varying systems with stochastic nonlinearities described as follows:

$$
\begin{aligned}
x(i+1,j+1) =\ & A_1(i+1,j)x(i+1,j) + A_2(i,j+1)x(i,j+1) \\
& + B_1(i+1,j)w(i+1,j) + B_2(i,j+1)w(i,j+1) \\
& + \alpha_1(i+1,j)f(x(i+1,j)) \\
& + \alpha_2(i,j+1)f(x(i,j+1)), \qquad i,j \in [0\ K] \quad (7.1)
\end{aligned}
$$

where $x(i,j) \in \mathbb{R}^{n_x}$ is the system state, $A_s(i,j)$ and $B_s(i,j)$ ($s = 1,2$) are known matrices with appropriate dimensions, $w(i,j) \in \mathbb{R}^p$ is the process noise obeying a Gaussian white distribution with $\mathbb{E}\{w(i,j)\} = 0$ and $\text{Var}\{w(i,j)\} = Q(i,j)$, and $\alpha_s(i,j) \in \mathbb{R}$ are random variables with $\mathbb{E}\{\alpha_s(i,j)\} = 0$ and $\text{Var}\{\alpha_s(i,j)\} = \sigma_{\alpha_s}$. The nonlinear function $f(\cdot)$ is to be defined later. For $i,j \in [0\ K]$, the initial conditions of (7.1) are given as

$$
\begin{aligned}
& \mathbb{E}\{x(i,0)\} = u_1(i), \qquad \mathbb{E}\{x(0,j)\} = u_2(j), \\
& \mathbb{E}\left\{ [x^T(i,0)\ x^T(0,j)]^T [x^T(t,0)\ x^T(0,r)] \right\} \\
& = \text{diag}\left\{ \delta(i,t)X(i,0), \delta(j,r)X(0,j) \right\},
\end{aligned}
$$

where $u_1(i)$ and $u_2(j)$ are known vectors subject to $u_1(0) = u_2(0) = 0$, $\delta(\cdot,\cdot)$ is the Kronecker delta function, $X(i,0)$ and $X(0,j)$ are the known second-order moments of the initial states $x(i,0)$ and $x(0,j)$, respectively.

For the sensor network consisting of m sensors, the communication topology is depicted by a directed graph $\mathcal{G} = (\mathcal{V}, \mathcal{E}, \mathcal{A})$ with the set of nodes $\mathcal{V} = \{1,2,\cdots,m\}$, the set of edges $\mathcal{E} \subseteq \mathcal{V} \times \mathcal{V}$, and the adjacency matrix $\mathcal{A} = (g_{lk})_{m \times m}$. The notation $(l,k) \in \mathcal{E}$ represents an edge starting at node k and ending at node l, namely the k-th node could send information to the l-th node and node k is thus regarded as a neighbor of node l. Elements in matrix $\mathcal{A}$ are nonnegative with $g_{lk} = 1 \Leftrightarrow (l,k) \in \mathcal{E}$, and $g_{lk} = 0$ otherwise. The neighbors of node l plus itself constitute the set $N_l \triangleq \{k \in \mathcal{V} | (l,k) \in \mathcal{E}\}$.

For $l \in [1\ m]$, the measurement of the l-th sensor is governed by

$$
y_l(i,j) = C_l(i,j)x(i,j) + \beta_l(i,j)h_l(x(i,j)) + v_l(i,j), \quad i,j \in [0\ K], \quad (7.2)
$$

where $y_l(i,j) \in \mathbb{R}^{n_y}$ is the measurement output received by the l-th sensor, $C_l(i,j)$ is the known shift-varying matrix, $v_l(i,j)$ is the measurement noises satisfying zero-mean Gaussian white distribution with $\text{Var}\{v_l(i,j)\} = R_l(i,j)$, and $\beta_l(i,j) \in \mathbb{R}$ is the random variable with $\mathbb{E}\{\beta_l(i,j)\} = 0$ and $\text{Var}\{\beta_l(i,j)\} = \sigma_{\beta_l}$.

The nonlinear functions $f : \mathbb{R}^{n_x} \to \mathbb{R}^{n_x}$ and $h_l : \mathbb{R}^{n_x} \to \mathbb{R}^{n_y}$ $(l \in [1\ m])$ are continuous and satisfy the following conditions:

$$f(0) = 0, \quad h_l(0) = 0, \tag{7.3a}$$
$$\|f(x) - f(z)\| \leq \mu\|x - z\|, \tag{7.3b}$$
$$\|h_l(x) - h_l(z)\| \leq \tau_l\|x - z\|, \tag{7.3c}$$

where x and z are any vectors belonging to $\mathbb{R}^{n_x}$, and μ and τ_l are known positive scalars.

Remark 7.1 *In industrial applications, nonlinear perturbations are usually generated which even occur in a random fashion. For example, during the dynamical evolutions of the wave equations competent for featuring the vibrating string issues and the heat equations available for depicting the water heating processes, the stochastic nonlinearities undoubtedly appear in view of the existence of elastic deformation, friction force, and damped oscillation under the randomly fluctuated situations. Additionally, nonlinear resistance in the circuit systems may also yield the nonlinear disturbances with certain knowledge of amplitude.*

7.1.2 Random Access Protocol

In what follows, the signal transmission method is introduced.

For technical analysis, represent the measurement of the l-th sensor (before being transmitted through the network) as

$$y_l(i,j) = [y_{l1}(i,j)\ y_{l2}(i,j) \cdots y_{ln_y}(i,j)]^T$$

i.e., the information received by the l-th sensor is first divided into n_y packets and then transmitted through the shared network. Note that the data may suffer from blockage or dropout due primarily to the constrained bandwidth and, therefore, the RA scheduling is introduced here to regulate the data transmission. To be more specific, for each sensor node, among the n_y packets of information, only one packet is allowed to be sent through the shared network at each transmission step.

For $l \in [1\ m]$, let $\xi_l(i,j) \in \{1, 2, \dots, n_y\}$ be the label index of the selected packet for the l-th sensor having access to the network at the shift step (i,j). Under the RA scheduling, $\xi_1(i,j), \xi_2(i,j), \dots, \xi_m(i,j)$ could be viewed as m mutually uncorrelated random variables [45]. The event $\xi_l(i,j) = k$ is assumed to satisfy the following probability distribution:

$$\text{Prob}\{\xi_l(i,j) = k\} = p_{lk}, \qquad k \in [1\ n_y], \tag{7.4}$$

where p_{lk} is a given positive scalar indicating the occurrence probability that the k-th signal/packet received by the l-th sensor node is chosen to be transmitted at the current transmission step. Obviously, the relationship $\sum_{k=1}^{n_y} p_{lk} = 1$ is satisfied for all $l \in [1\ m]$.

Denote the measured output of the l-th sensor after being transmitted as

$$\bar{y}_l(i,j) = [\bar{y}_{l1}(i,j) \ \bar{y}_{l2}(i,j) \cdots \bar{y}_{ln_y}(i,j)]^T.$$

Then the updating law for $\bar{y}_{lk}(i,j)$ under the RA protocol given above is derived as

$$\bar{y}_{lk}(i,j) = \begin{cases} y_{lk}(i,j), & \text{if } \xi_l(i,j) = k \\ 0, & \text{otherwise.} \end{cases} \tag{7.5}$$

A more compact form of $\bar{y}_l(i,j)$ can be presented as follows:

$$\bar{y}_l(i,j) = \Lambda_{\xi_l(i,j)} y_l(i,j), \tag{7.6}$$

where $\Lambda_{\xi_l(i,j)} \triangleq \mathrm{diag}_{1 \leq s \leq n_y} \{\delta(\xi_l(i,j),s)\}$.

The following assumptions are made in this chapter.

Assumption 7.1 *For $i,j \in [0 \ K]$ and $l \in [1 \ m]$, the random sequences $\xi_l(i,j)$ are mutually uncorrelated with all the other random variables, and so do the initial states $x(i,0)$ and $x(0,j)$.*

Assumption 7.2 *For $l,k \in [1 \ m]$, $i,j,t,r \in [0 \ K]$, and $s,\varsigma \in \{1,2\}$, the following relationship is satisfied:*

$$\mathbb{E}\left\{\eta_{sl}(i,j)\eta_{\varsigma k}^T(t,r)\right\}$$
$$= \mathrm{diag}\left\{Q(i,j), \delta(s,\varsigma)\sigma_{\alpha_s}, \delta(l,k)\sigma_{\beta_l}, \delta(l,k)R_l(i,j)\right\} \delta(i,t)\delta(j,r),$$

where

$$\eta_{sl}(i,j) \triangleq [w^T(i,j) \ \alpha_s(i,j) \ \beta_l(i,j) \ v_l^T(i,j)]^T.$$

7.1.3 Distributed Filter

In this chapter, a distributed filter structure is considered in order to utilize the local measurements from the m sensors to update the filters in a cooperative way.

For system (7.1), a distributed filter is proposed in the following form:

$$\hat{x}_l^-(i,j) = A_1(i,j-1)\hat{x}_l(i,j-1) + A_2(i-1,j)\hat{x}_l(i-1,j), \tag{7.7a}$$

$$\hat{x}_l(i,j) = \hat{x}_l^-(i,j) + \sum_{k \in N_l} g_{lk} K_{lk}(i,j)\hat{e}_k(i,j) \tag{7.7b}$$

with $\hat{e}_k(i,j) \triangleq \bar{y}_k(i,j) - \Lambda_{\xi_k(i,j)}C_k(i,j)\hat{x}_k^-(i,j)$ $(i,j \in [1 \ K])$, where $\hat{x}_l^-(i,j)$ and $\hat{x}_l(i,j) \in \mathbb{R}^{n_x}$ are the prediction and the estimation of $x(i,j)$ in the l-th sensor, respectively, and $K_{lk}(i,j)$ are the gain coefficients to be designed. The initial condition related to (7.7) is given as $\hat{x}_l(i,0) = u_1(i)$ and $\hat{x}_l(0,j) = u_2(j)$. The schematic diagram of the considered system with the distributed filtering scheme and the proposed protocol is shown in Fig. 7.1.

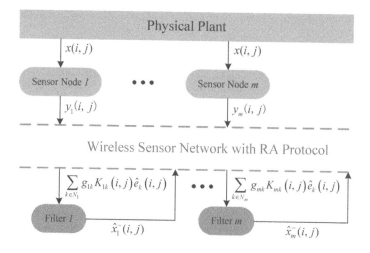

FIGURE 7.1: Configuration of the target system over the sensor network with RA protocol.

Remark 7.2 *The recursive distributed filter adopted here consists of the one-step prediction and estimation procedures as shown in (7.7a) and (7.7b), respectively. The innovation information $\sum_{k \in N_l} \hat{e}_k(i,j)$ covering the own measurement of sensor l and its neighbors will be exchanged through the sensor network under the RA protocol scheduling. In this case, the information collected from distributed nodes can be fully used to update the state estimate. Such a distributed filtering strategy has been preferred due to its clear physical implication and promising application prospect, see e.g., [118, 124].*

Define the prediction and estimation errors as

$$e_l^-(i,j) \triangleq x(i,j) - \hat{x}_l^-(i,j),$$
$$e_l(i,j) \triangleq x(i,j) - \hat{x}_l(i,j).$$

Then, the error dynamics is obtained from (7.1), (7.2), and (7.7) as follows:

$$
\begin{aligned}
e_l^-(i,j) = {} & A_1(i,j-1)e_l(i,j-1) + A_2(i-1,j)e_l(i-1,j) \\
& + B_1(i,j-1)w(i,j-1) + B_2(i-1,j)w(i-1,j) \\
& + \alpha_1(i,j-1)f(x(i,j-1)) + \alpha_2(i-1,j)f(x(i-1,j)), \quad (7.8a)
\end{aligned}
$$

$$
\begin{aligned}
e_l(i,j) = {} & e_l^-(i,j) - \sum_{k \in N_l} g_{lk} K_{lk}(i,j) \Lambda_{\xi_k(i,j)} \big[C_k(i,j) e_k^-(i,j) \\
& + \beta_k(i,j) h_k(x(i,j)) + v_k(i,j) \big]. \quad (7.8b)
\end{aligned}
$$

For brevity, we denote

$$e^-(i,j) \triangleq \mathrm{col}_m \left\{ e_l^-(i,j) \right\}, \quad e(i,j) \triangleq \mathrm{col}_m \left\{ e_l(i,j) \right\},$$

$$v(i,j) \triangleq \mathrm{col}_m \{v_l(i,j)\}, \quad R(i,j) \triangleq \mathrm{diag}_{1 \leq l \leq m} \{R_l(i,j)\},$$

$$K(i,j) \triangleq (K_{lk}(i,j))_{m \times m}, \quad h(x(i,j)) \triangleq \mathrm{col}_m \{h_l(x(i,j))\},$$

$$\bar{f}(x(i,j)) \triangleq \mathbf{1_m} \otimes f(x(i,j)), \quad \bar{A}_s(i,j) \triangleq I_m \otimes A_s(i,j),$$

$$\Lambda_{\xi(i,j)} \triangleq \mathrm{diag}_{1 \leq l \leq m} \{\Lambda_{\xi_l(i,j)}\}, \quad \bar{B}_s(i,j) \triangleq \mathbf{1_m} \otimes B_s(i,j),$$

$$C(i,j) \triangleq \mathrm{diag}_{1 \leq l \leq m} \{C_l(i,j)\}, \quad G_l \triangleq \mathrm{diag}_{1 \leq k \leq m} \{g_{lk}I\},$$

$$\Gamma_\beta(i,j) \triangleq \mathrm{diag}_{1 \leq l \leq m} \{\beta_l(i,j)I\}, \qquad s = 1,2$$

$$E_l \triangleq \mathrm{diag}\{\underbrace{0, \cdots, 0}_{l-1}, I, \underbrace{0, \cdots, 0}_{m-l}\}, \qquad l \in [1 \ m].$$

Then, a compact form for (7.8) is arranged as follows:

$$
\begin{aligned}
e^-(i,j) &= \bar{A}_1(i,j-1)e(i,j-1) + \bar{A}_2(i-1,j)e(i-1,j) \\
&\quad + \bar{B}_1(i,j-1)w(i,j-1) + \bar{B}_2(i-1,j)w(i-1,j) \\
&\quad + \alpha_1(i,j-1)\bar{f}(x(i,j-1)) + \alpha_2(i-1,j)\bar{f}(x(i-1,j)),
\end{aligned}
\tag{7.9a}
$$

$$
\begin{aligned}
e(i,j) &= e^-(i,j) - \sum_{l=1}^{m} E_l K(i,j) G_l \Lambda_{\xi(i,j)} \big[C(i,j)e^-(i,j) \\
&\quad + \Gamma_\beta(i,j)h(x(i,j)) + v(i,j) \big].
\end{aligned}
\tag{7.9b}
$$

Furthermore, by denoting $\bar{K}(i,j) = \sum_{l=1}^{m} E_l K(i,j) G_l$, one has

$$
\begin{aligned}
e(i,j) &= \big[I - \bar{K}(i,j)\Lambda_{\xi(i,j)}C(i,j) \big] e^-(i,j) \\
&\quad - \bar{K}(i,j)\Lambda_{\xi(i,j)} \big[\Gamma_\beta(i,j)h(x(i,j)) + v(i,j) \big].
\end{aligned}
\tag{7.10}
$$

The objective of the addressed problem is to design the filter gains $K_{lk}(i,j)$ for (7.1)–(7.2) such that the estimation error variance is bounded and, subsequently, the tightest upper bound (in the local and trace sense) is obtained over a prescribed finite horizon for all admissible RA scheduling.

7.2   Main Results

In this section, the existence of an upper bound is first ensured for the estimation error variance, and then the distributed filter is designed to locally optimize the developed upper bound at each iteration.

To begin with, we aim to establish an upper bound on the second-order moment of the system state as well as the dynamics of the filtering error variances. For simplicity, let us define

$$X(i,j) \triangleq \mathbb{E} \{x(i,j)x^T(i,j)\},$$

$$P^-(i,j) \triangleq \mathbb{E} \{e^-(i,j)(e^-(i,j))^T\},$$

$$P(i,j) \triangleq \mathbb{E}\left\{e(i,j)e^T(i,j)\right\}.$$

The following lemmas are provided which will be used in the subsequent developments.

Lemma 7.1 *For any given positive scalar λ, assume that there is a sequence of positive definite matrices $\{\bar{X}(i,j)\}_{i,j=1}^{K}$ such that the following recursion*

$$\begin{aligned}
\bar{X}(i,j) &= (1+\lambda)A_1(i,j-1)\bar{X}(i,j-1)A_1^T(i,j-1) \\
&\quad + (1+\lambda^{-1})A_2(i-1,j)\bar{X}(i-1,j)A_2^T(i-1,j) \\
&\quad + \mu^2\left(\sigma_{\alpha_1}\operatorname{tr}\left\{\bar{X}(i,j-1)\right\} + \sigma_{\alpha_2}\operatorname{tr}\left\{\bar{X}(i-1,j)\right\}\right)I \\
&\quad + B_1(i,j-1)Q(i,j-1)B_1^T(i,j-1) \\
&\quad + B_2(i-1,j)Q(i-1,j)B_2^T(i-1,j)
\end{aligned} \tag{7.11}$$

is satisfied with initial conditions $\bar{X}(i,0) = X(i,0)$ and $\bar{X}(0,j) = X(0,j)$ for $i,j \in [0\ K]$. Then, the second-order moment of the system state $X(i,j)$ is bounded by $\bar{X}(i,j)$.

Proof *Recalling conditions (7.3a)–(7.3b), one has*

$$\begin{aligned}
\mathbb{E}\left\{f(x(i,j))f^T(x(i,j))\right\} &\le \mathbb{E}\left\{f^T(x(i,j))f(x(i,j))\right\}I \\
&\le \mu^2\mathbb{E}\left\{x^T(i,j)x(i,j)\right\}I \\
&= \mu^2\operatorname{tr}\{X(i,j)\}I.
\end{aligned} \tag{7.12}$$

It follows from (7.1) and (7.12) that

$$\begin{aligned}
X(i,j) &= A_1(i,j-1)X(i,j-1)A_1^T(i,j-1) \\
&\quad + A_2(i-1,j)X(i-1,j)A_2^T(i-1,j) \\
&\quad + A_1(i,j-1)\mathbb{E}\{x(i,j-1)x^T(i-1,j)\}A_2^T(i-1,j) \\
&\quad + A_2(i-1,j)\mathbb{E}\{x(i-1,j)x^T(i,j-1)\}A_1^T(i,j-1) \\
&\quad + \sigma_{\alpha_1}\mathbb{E}\left\{f(x(i,j-1))f^T(x(i,j-1))\right\} \\
&\quad + \sigma_{\alpha_2}\mathbb{E}\left\{f(x(i-1,j))f^T(x(i-1,j))\right\} \\
&\quad + B_1(i,j-1)Q(i,j-1)B_1^T(i,j-1) \\
&\quad + B_2(i-1,j)Q(i-1,j)B_2^T(i-1,j) \\
&\le (1+\lambda)A_1(i,j-1)X(i,j-1)A_1^T(i,j-1) \\
&\quad + (1+\lambda^{-1})A_2(i-1,j)X(i-1,j)A_2^T(i-1,j) \\
&\quad + \mu^2\left(\sigma_{\alpha_1}\operatorname{tr}\{X(i,j-1)\} + \sigma_{\alpha_2}\operatorname{tr}\{X(i-1,j)\}\right)I \\
&\quad + B_1(i,j-1)Q(i,j-1)B_1^T(i,j-1) \\
&\quad + B_2(i-1,j)Q(i-1,j)B_2^T(i-1,j).
\end{aligned} \tag{7.13}$$

Subtracting (7.11) from (7.13) yields

$$
\begin{aligned}
X(i,j) &- \bar{X}(i,j) \\
&\leq (1+\lambda)A_1(i,j-1)\left(X(i,j-1) - \bar{X}(i,j-1)\right)A_1^T(i,j-1) \\
&\quad + (1+\lambda^{-1})A_2(i-1,j)\left(X(i-1,j) - \bar{X}(i-1,j)\right)A_2^T(i-1,j) \\
&\quad + \mu^2\sigma_{\alpha_1}\left(\mathrm{tr}\left\{X(i,j-1)\right\} - \mathrm{tr}\left\{\bar{X}(i,j-1)\right\}\right)I \\
&\quad + \mu^2\sigma_{\alpha_2}\left(\mathrm{tr}\left\{X(i-1,j)\right\} - \mathrm{tr}\left\{\bar{X}(i-1,j)\right\}\right)I.
\end{aligned}
\tag{7.14}
$$

Note that the initial conditions $\bar{X}(i,0) = X(i,0)$ and $\bar{X}(0,j) = X(0,j)$ indicate the validity of $X(i,j) \leq \bar{X}(i,j)$ for $(i,j) \in \{(i_0,j_0)|i_0,j_0 \geq 0; i_0 + j_0 = 1\}$. Furthermore, assuming that $X(i,j) \leq \bar{X}(i,j)$ is true for $(i,j) \in \{(i_0,j_0)|i_0,j_0 \geq 0; i_0 + j_0 = k\}$, it is not difficult to derive from the trace property and (7.14) that $X(i,j) \leq \bar{X}(i,j)$ holds for $(i,j) \in \{(i_0,j_0)|i_0,j_0 > 0; i_0 + j_0 = k+1\}$. The proof is hence verified from mathematical induction.

Lemma 7.2 *The recursion of the prediction error variance is given as follows:*

$$
\begin{aligned}
P^-(i,j) &= \bar{A}_1(i,j-1)P(i,j-1)\bar{A}_1^T(i,j-1) \\
&\quad + \bar{A}_2(i-1,j)P(i-1,j)\bar{A}_2^T(i-1,j) \\
&\quad + \bar{A}_1(i,j-1)\mathbb{E}\{e(i,j-1)e^T(i-1,j)\}\bar{A}_2^T(i-1,j) \\
&\quad + \bar{A}_2(i-1,j)\mathbb{E}\{e(i-1,j)e^T(i,j-1)\}\bar{A}_1^T(i,j-1) \\
&\quad + \sigma_{\alpha_1}\mathbb{E}\left\{\bar{f}(x(i,j-1))\bar{f}^T(x(i,j-1))\right\} \\
&\quad + \sigma_{\alpha_2}\mathbb{E}\left\{\bar{f}(x(i-1,j))\bar{f}^T(x(i-1,j))\right\} \\
&\quad + \bar{B}_1(i,j-1)Q(i,j-1)\bar{B}_1^T(i,j-1) \\
&\quad + \bar{B}_2(i-1,j)Q(i-1,j)\bar{B}_2^T(i-1,j).
\end{aligned}
\tag{7.15}
$$

Proof *Owing to the statistical properties of random variables $w(i,j)$ and $\alpha_s(i,j)$ $(s = 1,2)$, the proof of this lemma follows immediately.*

Lemma 7.3 *The recursion of the estimation error variance is given as follows:*

$$
\begin{aligned}
P(i,j) &= \mathbb{E}\left\{[I - \bar{K}(i,j)\Lambda_{\xi(i,j)}C(i,j)]P^-(i,j)[I - \bar{K}(i,j)\Lambda_{\xi(i,j)}C(i,j)]^T\right\} \\
&\quad + \bar{K}(i,j)\mathbb{E}\left\{\Lambda_{\xi(i,j)}[\Gamma_\beta(i,j)h(x(i,j))h^T(x(i,j)) \right. \\
&\quad \left. \times \Gamma_\beta(i,j) + R(i,j)]\Lambda_{\xi(i,j)}\right\}\bar{K}^T(i,j).
\end{aligned}
\tag{7.16}
$$

Proof *Note that the random variables $e^-(i,j)$, $\Gamma_\beta(i,j)$, and $v(i,j)$ are uncorrelated with each other for all $i,j \in [0\ K]$. The assertion of this lemma can be obtained easily and hence the proof is omitted for brevity.*

Remark 7.3 *So far, the dynamics of the filtering error variances has been* *presented. In view of the nonlinearities involved in the error dynamics, the analytical solution of the estimation error variance $P(i,j)$ cannot be obtained, needless to say the design of the filter gain parameter. To this end, an alternative method is proposed to tackle the distributed filtering problem by constructing the tightest upper bound of $P(i,j)$ at each shift step.*

Before presenting the main results, some statistical properties of the stochastic matrix $\Lambda_{\xi(i,j)}$ are carefully analysed. By defining $\Pi_l \triangleq \mathrm{diag}_{1\leq t\leq n_y}\{p_{lt}\} \in \mathbb{R}^{n_y\times n_y}$, one obtains from the expressions of $\Lambda_{\xi_l(i,j)}$ and $\Lambda_{\xi(i,j)}$ that

$$\mathbb{E}\{\Lambda_{\xi_l(i,j)}\} = \Pi_l, \quad \mathbb{E}\{\Lambda_{\xi(i,j)}\} = \mathrm{diag}_{1\leq l\leq m}\{\Pi_l\} \triangleq \bar{\Pi}.$$

Based on the definition of $\delta(\cdot,\cdot)$ and the probability distributions of $\xi_l(i,j)$, for $k,l \in [1\ m]$ and $s,t \in [1\ n_y]$, it is evident to have the following equalities:

$$\mathbb{E}\left\{\delta^2(\xi_k(i,j),s)\right\} = p_{ks},$$
$$\mathbb{E}\left\{\delta(\xi_k(i,j),s)\delta(\xi_k(i,j),t)\right\} = 0, \quad s \neq t$$
$$\mathbb{E}\left\{\delta(\xi_k(i,j),s)\delta(\xi_l(i,j),t)\right\} = p_{ks}p_{lt}, \quad k \neq l.$$

Then, by resorting to matrix analysis, one confirms that

$$\mathbb{E}\{\Lambda_{\xi(i,j)}Z\Lambda_{\xi(i,j)}^T\} = \hat{\Pi} \circ Z, \qquad (7.17)$$
$$\tilde{\Pi} \circ Z = \hat{\Pi} \circ Z - \bar{\Pi}Z\bar{\Pi}^T, \qquad (7.18)$$

where Z is a given matrix with compatible dimensions, $\hat{\Pi} = (\hat{\Pi}_{kl})_{m\times m} \in \mathbb{R}^{mn_y\times mn_y}$ satisfying

$$\hat{\Pi}_{kl} = \begin{cases} \Pi_k, & k = l \\ \Pi_k \mathbf{1}\mathbf{1}^T\Pi_l, & k \neq l \end{cases}$$

and $\tilde{\Pi} = \mathrm{diag}_{1\leq l\leq m}\{\tilde{\Pi}_l\}$ with $\tilde{\Pi}_l = (\tilde{\Pi}_{l,st})_{n_y\times n_y}$

$$\tilde{\Pi}_{l,st} = \begin{cases} p_{ls}(1 - p_{ls}), & s = t \\ -p_{ls}p_{lt}, & s \neq t. \end{cases}$$

We are now in the position to ascertain the existence of an upper bound on the estimation error variance with rigorous derivations.

Theorem 7.1 *For given positive scalars λ and ξ, assume that there are two sequences of positive matrices $\{M^-(i,j)\}_{i,j=1}^K$ and $\{M(i,j)\}_{i,j=1}^K$ satisfying the following equations*

$$M^-(i,j) = (1+\lambda)\bar{A}_1(i,j-1)M(i,j-1)\bar{A}_1^T(i,j-1)$$
$$+ (1+\lambda^{-1})\bar{A}_2(i-1,j)M(i-1,j)\bar{A}_2^T(i-1,j)$$
$$+ \left(\sigma_{\alpha_1}\mathrm{tr}\{\bar{X}(i,j-1)\} + \sigma_{\alpha_2}m\mu^2\mathrm{tr}\left\{\bar{X}(i-1,j)\right\}\right)I$$

$$+ \bar{B}_1(i, j-1)Q(i, j-1)\bar{B}_1^T(i, j-1)$$
$$+ \bar{B}_2(i-1, j)Q(i-1, j)\bar{B}_2^T(i-1, j), \tag{7.19}$$

$$M(i, j) = \left(I - \bar{K}(i, j)\bar{\Pi}C(i, j)\right) M^-(i, j) \left(I - \bar{K}(i, j)\bar{\Pi}C(i, j)\right)^T$$
$$+ \bar{K}(i, j)\left[\tilde{\Pi} \circ \left(C(i, j)M^-(i, j)C^T(i, j)\right)\right]\bar{K}^T(i, j)$$
$$+ \bar{K}(i, j)\left[\hat{\Pi} \circ \bar{R}(i, j) + \xi I\right]\bar{K}^T(i, j) \tag{7.20}$$

with initial constraints

$$M(i, 0) = P(i, 0), \quad M(0, j) = P(0, j) \tag{7.21}$$

for $i, j \in [0 \ K]$, where

$$\bar{R}(i, j) \triangleq \max_{1 \le l \le m} \{\sigma_{\beta_l}\} \sum_{l=1}^{m} \tau_l^2 \mathrm{tr}\{\bar{X}(i, j)\}I + R(i, j).$$

Then, the error variances concerning system (7.1) are bounded by

$$P^-(i, j) \le M^-(i, j), \quad P(i, j) \le M(i, j). \tag{7.22}$$

Proof *In order to prove the validity of this theorem, the foremost procedure is to further analyze recursions of the error variances. According to (7.3a), (7.3b) and Lemma 7.1, one has*

$$\mathbb{E}\left\{\bar{f}(x(i, j))\bar{f}^T(x(i, j))\right\} \le \mathbb{E}\left\{\bar{f}^T(x(i, j))\bar{f}(x(i, j))\right\} I_{mn_x}$$
$$\le m\mu^2 \mathbb{E}\left\{x^T(i, j)x(i, j)\right\} I_{mn_x}$$
$$\le m\mu^2 \mathrm{tr}\left\{\bar{X}(i, j)\right\} I_{mn_x}. \tag{7.23}$$

Similarly, it follows from (7.3a) and (7.3c) that

$$\mathbb{E}\left\{h(x(i, j))h^T(x(i, j))\right\} \le \sum_{l=1}^{m} \tau_l^2 \mathrm{tr}\{\bar{X}(i, j)\}I_{mn_y}. \tag{7.24}$$

Substituting (7.23) into (7.15) yields

$$P^-(i, j) \le (1 + \lambda)\bar{A}_1(i, j-1)P(i, j-1)\bar{A}_1^T(i, j-1)$$
$$+ (1 + \lambda^{-1})\bar{A}_2(i-1, j)P(i-1, j)\bar{A}_2^T(i-1, j)$$
$$+ m\mu^2 \left(\sigma_{\alpha_1}\mathrm{tr}\{\bar{X}(i, j-1)\} + \sigma_{\alpha_2}\mathrm{tr}\{\bar{X}(i-1, j)\}\right) I$$
$$+ \bar{B}_1(i, j-1)Q(i, j-1)\bar{B}_1^T(i, j-1)$$
$$+ \bar{B}_2(i-1, j)Q(i-1, j)\bar{B}_2^T(i-1, j). \tag{7.25}$$

By applying the statistical property of $\Gamma_\beta(i, j)$ and the uncorrelatedness among $x(i, j)$, $\Gamma_\beta(i, j)$, and $\Lambda_{\xi(i,j)}$, we derive the following inequality (7.26) where (7.17) and (7.24) have also been utilized in the derivation:

$$\mathbb{E}\left\{\Lambda_{\xi(i,j)}\left[\Gamma_\beta(i, j)h(x(i, j))h^T(x(i, j))\Gamma_\beta^T(i, j) + R(i, j)\right]\Lambda_{\xi(i,j)}^T\right\}$$

$$= \mathbb{E}\left\{\Lambda_{\xi(i,j)}\left[\mathbb{E}\left\{\Gamma_\beta(i,j)\mathbb{E}\{h(x(i,j))h^T(x(i,j))\}\Gamma_\beta^T(i,j)\right\} + R(i,j)\right]\Lambda_{\xi(i,j)}^T\right\}$$

$$\leq \mathbb{E}\left\{\Lambda_{\xi(i,j)}\left[\sum_{l=1}^{m}\tau_l^2\text{tr}\{\bar{X}(i,j)\}\mathbb{E}\{\Gamma_\beta(i,j)\Gamma_\beta^T(i,j)\} + R(i,j)\right]\Lambda_{\xi(i,j)}^T\right\}$$

$$\leq \hat{\Pi}\circ\left[\max_{1\leq l\leq m}\{\sigma_{\beta_l}\}\sum_{l=1}^{m}\tau_l^2\text{tr}\{\bar{X}(i,j)\}I + R(i,j)\right]. \tag{7.26}$$

It is clear to see from (7.16) to (7.18) and (7.26) that

$$\begin{aligned}
P(i,j) &\leq P^-(i,j) - \bar{K}(i,j)\bar{\Pi}C(i,j)P^-(i,j) - P^-(i,j)C^T(i,j)\bar{\Pi}^T\bar{K}^T(i,j)\\
&\quad + \bar{K}(i,j)\left[\hat{\Pi}\circ\left(C(i,j)P^-(i,j)C^T(i,j) + \bar{R}(i,j)\right)\right]\bar{K}^T(i,j)\\
&= \left(I - \bar{K}(i,j)\bar{\Pi}C(i,j)\right)P^-(i,j)\left(I - \bar{K}(i,j)\bar{\Pi}C(i,j)\right)^T\\
&\quad + \bar{K}(i,j)\left[\hat{\Pi}\circ\bar{R}(i,j) + \hat{\Pi}\circ\left(C(i,j)P^-(i,j)C^T(i,j)\right)\right.\\
&\quad \left. - \bar{\Pi}\left(C(i,j)P^-(i,j)C^T(i,j)\right)\bar{\Pi}^T\right]\bar{K}^T(i,j)\\
&= \left(I - \bar{K}(i,j)\bar{\Pi}C(i,j)\right)P^-(i,j)\left(I - \bar{K}(i,j)\bar{\Pi}C(i,j)\right)^T\\
&\quad + \bar{K}(i,j)\left[\hat{\Pi}\circ\bar{R}(i,j) + \tilde{\Pi}\circ\left(C(i,j)P^-(i,j)C^T(i,j)\right)\right]\bar{K}^T(i,j).
\end{aligned} \tag{7.27}$$

Now, it is ready to confirm the assertion of (7.22) by the inductive method. It follows from (7.19), (7.25), and the initial condition (7.21) that

$$\begin{aligned}
&P^-(1,1) - M^-(1,1)\\
&\leq (1+\lambda)\bar{A}_1(1,0)(P(1,0) - M(1,0)\bar{A}_1^T(1,0)\\
&\quad + (1+\lambda^{-1})\bar{A}_2(0,1)(P(0,1) - M(0,1))\bar{A}_2^T(0,1)\\
&\leq 0
\end{aligned}$$

which, together with (7.20), (7.27), and the algebra matrix knowledge, leads to

$$\begin{aligned}
&P(1,1) - M(1,1)\\
&\leq \left(I - \bar{K}(1,1)\bar{\Pi}C(1,1)\right)\left(P^-(1,1) - M^-(1,1)\right)\left(I - \bar{K}(1,1)\bar{\Pi}C(1,1)\right)^T\\
&\quad + \bar{K}(1,1)\left[\tilde{\Pi}\circ\left(C(1,1)(P^-(1,1) - M^-(1,1))C^T(1,1)\right) - \xi I\right]\bar{K}^T(1,1)\\
&\leq 0.
\end{aligned}$$

Hence, inequality (7.22) is true for $(i,j) \in \{(i_0,j_0)|i_0,j_0 > 0; i_0 + j_0 = 2\}$.

For a given integer $k \in [2\ \ 2K-1]$, assume that (7.22) holds for $(i,j) \in \{(i_0,j_0)|i_0,j_0 > 0; i_0 + j_0 = k\}$. Then, it follows from (7.19) to (7.20), (7.25), and (7.27) that

$$P^-(i,j) - M^-(i,j)$$

$$\leq (1+\lambda)\bar{A}_1(i,j-1)(P(i,j-1) - M(i,j-1)\bar{A}_1^T(i,j-1)$$
$$+ (1+\lambda^{-1})\bar{A}_2(i-1,j)(P(i-1,j) - M(i-1,j))\bar{A}_2^T(i-1,j)$$
$$\leq 0$$

is valid for $(i,j) \in \{(i_0,j_0)|i_0,j_0 > 0; i_0 + j_0 = k+1\}$, *which immediately yields*

$$P(i,j) - M(i,j)$$
$$\leq \left(I - \bar{K}(i,j)\bar{\Pi}C(i,j)\right)\left(P^-(i,j) - M^-(i,j)\right)\left(I - \bar{K}(i,j)\bar{\Pi}C(i,j)\right)^T$$
$$+ \bar{K}(i,j)\left[\tilde{\Pi} \circ \left(C(i,j)(P^-(i,j) - M^-(i,j))C^T(i,j)\right) - \xi I\right]\bar{K}^T(i,j)$$
$$\leq 0.$$

The proof of this theorem is hence concluded by induction.

Remark 7.4 *Based on the established preliminary results, an upper bound is provided in Theorem 7.1 for the estimation error variance which can be calculated by iteratively solving two sets of Riccati-like equations (7.19) and (7.20). Owing to the possible nonuniqueness of solutions to (7.19) and (7.20), the locally minimal upper bound $M(i,j)$ shall be determined to optimize the filtering performance by properly designing the distributed filter gains.*

In what follows, the distributed filter design strategy is analyzed such that the obtained upper bound $M(i,j)$ is minimized at each shift step. Before proceeding, we write matrix G_l ($l \in [1\ m]$) as $G_l \triangleq \bar{G}_l\bar{G}_l^T$ with $\bar{G}_l = \text{diag}_{1\leq k\leq m}\{\sqrt{g_{lk}}I\}$. Noting that $g_{lk} = 0$ ($k \notin N_l$) leads to the sparsity of matrix $\bar{G}_l$, one has the simplified matrix $\hat{G}_l$ with $G_l = \hat{G}_l\hat{G}_l^T$ by removing the corresponding zero columns from $\bar{G}_l$.

For notional simplicity, we denote

$$K(i,j) \triangleq \left[\ K_1^T(i,j)\ \ K_2^T(i,j)\ \ \cdots\ \ K_m^T(i,j)\ \right]^T,$$
$$M^-(i,j) \triangleq \left[\ M_1^{-T}(i,j)\ \ M_2^{-T}(i,j)\ \ \cdots\ \ M_m^{-T}(i,j)\ \right]^T,$$
$$\hat{R}(i,j) \triangleq \hat{\Pi} \circ \left(C(i,j)M^-(i,j)C^T(i,j) + \bar{R}(i,j)\right) + \xi I$$

and, for $l \in [1\ m]$,

$$K_l(i,j) \triangleq \left[\ K_{l1}(i,j)\ \ K_{l2}(i,j)\ \ \cdots\ \ K_{lm}(i,j)\ \right],$$
$$M_l^-(i,j) \triangleq \left[\ M_{l1}^-(i,j)\ \ M_{l2}^-(i,j)\ \ \cdots\ \ M_{lm}^-(i,j)\ \right],$$
$$\Gamma_l(i,j) \triangleq M_l^-(i,j)C^T(i,j)\bar{\Pi}\hat{G}_l\left(\hat{G}_l^T\hat{R}(i,j)\hat{G}_l\right)^{-1}\hat{G}_l^T$$
$$\triangleq \left[\ \Gamma_{l1}(i,j)\ \ \Gamma_{l2}(i,j)\ \ \cdots\ \ \Gamma_{lm}(i,j)\ \right].$$

The design of the distributed filter gain is presented in the following theorem.

Theorem 7.2 *For the estimation error variance concerning system (7.1), the obtained upper bound $M(i,j)$ is locally minimized in the trace sense by determining the filter gain as follows:*

$$K_{lk}(i,j) = \begin{cases} \Gamma_{lk}(i,j), & g_{lk} = 1 \\ 0, & g_{lk} = 0. \end{cases} \tag{7.28}$$

Proof *It follows from (7.20) that*

$$\begin{aligned}
M(i,j) &= M^-(i,j) - \bar{K}(i,j)\bar{\Pi}C(i,j)M^-(i,j) \\
&\quad - M^-(i,j)C^T(i,j)\bar{\Pi}^T\bar{K}^T(i,j) \\
&\quad + \bar{K}(i,j)\Big[\hat{\Pi} \circ \bar{R}(i,j) + \tilde{\Pi} \circ \big(C(i,j)M^-(i,j)C^T(i,j)\big) \\
&\quad + \bar{\Pi}C(i,j)M^-(i,j)C^T(i,j)\bar{\Pi}^T + \xi I\Big]\bar{K}^T(i,j) \\
&= M^-(i,j) - \bar{K}(i,j)\bar{\Pi}C(i,j)M^-(i,j) \\
&\quad - M^-(i,j)C^T(i,j)\bar{\Pi}^T\bar{K}^T(i,j) + \bar{K}(i,j)\hat{R}(i,j)\bar{K}^T(i,j) \tag{7.29}
\end{aligned}$$

where (7.18) has been applied in the second step of the above derivation. Taking the trace for both sides of (7.29) yields

$$\begin{aligned}
\mathrm{tr}\{M(i,j)\} = \mathrm{tr}\Big\{&M^-(i,j) - 2\bar{K}(i,j)\bar{\Pi}C(i,j)M^-(i,j) \\
&+ \bar{K}(i,j)\hat{R}(i,j)\bar{K}^T(i,j)\Big\}.
\end{aligned}$$

According to the definition of E_l ($l \in [1\ m]$) and the trace property, one has

$$\mathrm{tr}\left\{E_l Z E_k^T\right\} = 0, \quad l \neq k$$

where Z is an arbitrary matrix with appropriate dimensions. As a result, it is easy to derive that

$$\begin{aligned}
&\mathrm{tr}\{\bar{K}(i,j)\hat{R}(i,j)\bar{K}^T(i,j)\} \\
&= \mathrm{tr}\Big\{\Big(\sum_{l=1}^{m}E_l K(i,j)G_l\Big)\hat{R}(i,j)\Big(\sum_{l=1}^{m}E_l K(i,j)G_l\Big)^T\Big\} \\
&= \mathrm{tr}\Big\{\sum_{l=1}^{m}E_l K(i,j)G_l\hat{R}(i,j)G_l^T K^T(i,j)E_l^T\Big\}.
\end{aligned}$$

Taking the partial derivation of $\mathrm{tr}\{M(i,j)\}$ with respect to $K(i,j)$, one has the following equality

$$\frac{\partial \mathrm{tr}\{M(i,j)\}}{\partial K(i,j)} = -2\sum_{l=1}^{m}E_l^T M^-(i,j)C^T(i,j)\bar{\Pi}G_l^T$$

$$+ 2 \sum_{l=1}^{m} E_l^T E_l K(i,j) G_l \hat{R}(i,j) G_l^T.$$

To minimize $\text{tr}\{M(i,j)\}$, *the above partial derivation is set as zero, and this gives rise to*

$$\sum_{l=1}^{m} E_l K(i,j) G_l \hat{R}(i,j) G_l^T = \sum_{l=1}^{m} E_l M^-(i,j) C^T(i,j) \bar{\Pi} G_l^T$$

which is equivalent to

$$K_l(i,j) G_l \hat{R}(i,j) G_l^T = M_l^-(i,j) C^T(i,j) \bar{\Pi} G_l^T \qquad (7.30)$$

for $l \in [1\ m]$. *In view of* $G_l = \hat{G}_l \hat{G}_l^T$, *equation (7.30) is rewritten as*

$$K_l(i,j) \hat{G}_l \hat{G}_l^T \hat{R}(i,j) \hat{G}_l \hat{G}_l^T = M_l^-(i,j) C^T(i,j) \bar{\Pi} \hat{G}_l \hat{G}_l^T.$$

Recalling that matrix $\hat{G}_l^T$ *is full-row rank (and hence right invertible), one has*

$$K_l(i,j) \hat{G}_l \hat{G}_l^T \hat{R}(i,j) \hat{G}_l = M_l^-(i,j) C^T(i,j) \bar{\Pi} \hat{G}_l.$$

In addition, it is not difficult to verify the invertibility of $\hat{G}_l^T \hat{R}(i,j) \hat{G}_l$ ($l \in [1\ m]$) *from the matrix theory, which implies that*

$$K_l(i,j) \hat{G}_l = M_l^-(i,j) C^T(i,j) \bar{\Pi} \hat{G}_l \big(\hat{G}_l^T \hat{R}(i,j) \hat{G}_l\big)^{-1}.$$

Then, one has the following equality

$$K_l(i,j) G_l = M_l^-(i,j) C^T(i,j) \bar{\Pi} \hat{G}_l \big(\hat{G}_l^T \hat{R}(i,j) \hat{G}_l\big)^{-1} \hat{G}_l^T$$

which indicates

$$K_{lk}(i,j) = \Gamma_{lk}(i,j) \quad \text{when} \quad g_{lk} = 1.$$

Furthermore, $K_{lk}(i,j)$ *is selected as* $K_{lk}(i,j) = 0$ *for* $k \notin N_l$ *since there is no information sent from the non-neighbor nodes to the local sensor. Consequently, the gain parameter* $K_{lk}(i,j)$ *is determined by (7.28), which ends the proof.*

Remark 7.5 *The distributed filter is designed in Theorem 7.2 to guarantee the locally minimal upper bound* $M(i,j)$ *(in the trace sense) by fully using the topology information. A remarkable challenging encountered is that the gain parameter cannot be directly obtained from (7.30) on account of the sparsity of the network topology. Particularly, since* $g_{lk} = 1$ *if and only if* $k \in N_l$, *matrix* G_l *would be rank deficient and thus* $G_l \hat{R}(i,j) G_l^T$ *is likely to be singular. In this case, the matrix simplification technique is used to deal with* *the sparse sensor network and ascertain the nonsingularity of the simplified matrix $\hat{G}_l^T \hat{R}(i,j) \hat{G}_l$. Such a method has been first introduced in [119] (where the event-based distributed filtering problem has been investigated) and then also exploited in some recent literature on the filtering design issues with sensor networks, see e.g., [219].*

Remark 7.6 *It is worth mentioning that the attempt we have made in this chapter represents one of the first few on the study of the recursive distributed filtering problem for 2-D shift-varying systems. An upper bound of the filtering error variance is constructed in terms of Riccati-like equations, and then the filter parameters that minimize the obtained upper bound are designed in a distributed setting. Compared with the 2-D filtering/estimation issues in [269], the consideration of the RA scheduling and the possibly sparse topology contributes to the main difficulties in designing the proposed filtering scheme. It can be seen from the obtained results that the established upper bound and the designed filter gain are closely related to the communication protocol and the topology information.*

7.3 Numerical Example

In this section, the proposed distributed filtering strategy is applied to an industrial heating process to show the efficiency and applicability of the developed results. Such a process can be described by the following differential equation [83]:

$$\frac{\partial T(z,t)}{\partial t} + \frac{\partial T(z,t)}{\partial z} = aT(z,t) + bf(z,t),$$

where $T(z,t)$ is the reactor temperature to be estimated over the space $z \in [0, z_0]$ and time $t \in [0, t_0]$, $f(z,t)$ is a certain force function, parameters a and b are known scalars indicating the heat transfer coefficients.

Similar to the manipulations presented in the previous chapter, let's denote $T(z,t) \triangleq T_d(i\Delta z, j\Delta t)$ for $z \in [i\Delta z, (i+1)\Delta z)$ and $t \in [j\Delta t, (j+1)\Delta t)$, and then the following approximate relationships are obtained by properly choosing step sizes Δz and Δt:

$$\frac{\partial T(z,t)}{\partial t} \doteq \frac{u_d(i\Delta z, (j+1)\Delta t) - u_d(i\Delta z, j\Delta t)}{\Delta t},$$

$$\frac{\partial T(z,t)}{\partial z} \doteq \frac{u_d(i\Delta z, j\Delta t) - u_d((i-1)\Delta z, j\Delta t)}{\Delta z}.$$

Further set $\bar{T}(i,j) \triangleq T_d(i\Delta z, j\Delta t)$ and $x(i,j) \triangleq [\bar{T}^T(i-1,j) \ \bar{T}^T(i,j)]^T$. In the case of $b = 0$, the discretized model can be approximately expressed by the following 2-D system:

$$x(i+1, j+1) = A_1 x(i+1, j) + A_2 x(i, j+1)$$
$$+ \alpha_1(i+1, j) f(x(i+1, j)) + B_1 w(i+1, j)$$
$$+ \alpha_2(i, j+1) f(x(i, j+1)) + B_2 w(i, j+1)$$

with

$$A_1 = \begin{bmatrix} 0 & 0 \\ \frac{\Delta t}{\Delta z} & 1 - \frac{\Delta t}{\Delta z} + a\Delta t \end{bmatrix}, \quad A_2 = \begin{bmatrix} 0 & 1 \\ 0 & 0 \end{bmatrix}.$$

Here, the noise $w(i, j)$ is ineluctable due to the contaminated chemical reactor in practice. The stochastic nonlinearity $f(\cdot)$ is likely to emerge from the possible elastic deformation, friction force and/or damped oscillation, whose occurrence is determined by the random variable $\alpha_s(i, j)$ ($s = 1, 2$). The parameter matrices A_s and B_s would be shift-varying as the underlying system may suffer from heterogeneous media induced by the changeable environment. In the regard, the following shift-varying system parameters are considered:

$$\Delta t = 0.06, \quad \Delta z = 0.2, \quad a(i, j) = -2.5 \sin(i) \cos(j) - 5,$$

$$A_1(i, j) = \begin{bmatrix} 0 & 0 \\ 0.3 - 0.1e^{-2i} & 0.4 - 0.15 \sin(i) \cos(j) \end{bmatrix},$$

$$A_2(i, j) = \begin{bmatrix} 0 & 1 - 0.12 \sin(j) \\ 0 & 0 \end{bmatrix}, \quad K = 60,$$

$$B_1(i, j) = \begin{bmatrix} -0.05 \\ 0.1 + 0.1e^{-2i} \end{bmatrix}, \quad B_2(i, j) = \begin{bmatrix} 0.14 \\ -0.02 \cos(j) \end{bmatrix}.$$

To enhance the filtering performance of the considered system, a sensor network with ten smart nodes is utilized in a collaboratively deployed fashion to measure/monitor the temperature. The communication topology is described by a directed graph $\mathcal{G}$ with $\mathcal{V} = \{1, 2, 3, 4, 5, 6, 7, 8, 9, 10\}$ and $\mathcal{E} = \{(1, 4), (2, 5), (3, 6), (3, 10), (4, 7), (5, 3), (6, 9), (7, 6), (8, 4), (9, 3), (10, 1)\} \cup \{(l, l)| l \in [1, 10]\}$. The measurement model of the l-th sensor is given by

$$y_l(i, j) = C_l(i, j) x(i, j) + \beta_l(i, j) h_l(x(i, j)) + v_l(i, j)$$

where the measurement is sent through a shared channel to the remote filter with exchanged information among neighbors for deriving the local estimate. The output information is subject to the measured noise $v_l(i, j)$ and the nonlinear perturbation $h_l(\cdot)$ induced by undesired working conditions. Moreover, the RA protocol is adopted to schedule the transmission priority of signals for each sensor. The corresponding parameters are chosen as

$$C_1(i, j) = \begin{bmatrix} 0.3 & 0.5 \\ 0.6 + 0.1e^{-3i} & 1 \end{bmatrix}, \quad C_2(i, j) = \begin{bmatrix} -0.3 & 0.15 \sin(i) \\ 0.35 & 0.2 \end{bmatrix},$$

$$C_3(i, j) = \begin{bmatrix} 0.45 & 0.1 \\ 0.1 \cos(2j) & -0.15 \end{bmatrix}, \quad C_5(i, j) = \begin{bmatrix} 0.7 + 0.1e^{-3i} & 0.5 \\ -0.68 & 1 \end{bmatrix},$$

$$C_4(i,j) = \begin{bmatrix} 0.8 & 0 \\ 0.4 & 0 \end{bmatrix}, \quad C_6(i,j) = \begin{bmatrix} 0.25 & -0.4 \\ 0.35 - 0.2\sin(j) & 0.35 \end{bmatrix},$$

$$C_7(i,j) = \begin{bmatrix} -0.6 & 0.1\cos(2j) \\ 0.75 & 0.15 \end{bmatrix}, \quad C_8(i,j) = \begin{bmatrix} 0.8 & 0.7 \\ 0 & 0.3\cos(i) \end{bmatrix},$$

$$C_9(i,j) = \begin{bmatrix} 0.4 + 0.1\sin(3i) & 0.1 \\ 0.35 & 0.7 \end{bmatrix}, \quad C_{10}(i,j) = \begin{bmatrix} 0.2 & 1 \\ 0.45 & 0.6 \end{bmatrix}.$$

The probability distributions associated with the RA protocol are set to be

$$\text{prob}\{\xi_1(i,j) = 1\} = 0.4, \quad \text{prob}\{\xi_1(i,j) = 2\} = 0.6,$$
$$\text{prob}\{\xi_2(i,j) = 1\} = 0.5, \quad \text{prob}\{\xi_2(i,j) = 2\} = 0.5,$$
$$\text{prob}\{\xi_3(i,j) = 1\} = 0.6, \quad \text{prob}\{\xi_3(i,j) = 2\} = 0.4,$$
$$\text{prob}\{\xi_4(i,j) = 1\} = 0.3, \quad \text{prob}\{\xi_4(i,j) = 2\} = 0.7,$$
$$\text{prob}\{\xi_5(i,j) = 1\} = 0.2, \quad \text{prob}\{\xi_5(i,j) = 2\} = 0.8,$$
$$\text{prob}\{\xi_6(i,j) = 1\} = 0.45, \quad \text{prob}\{\xi_6(i,j) = 2\} = 0.55,$$
$$\text{prob}\{\xi_7(i,j) = 1\} = 0.55, \quad \text{prob}\{\xi_7(i,j) = 2\} = 0.45,$$
$$\text{prob}\{\xi_8(i,j) = 1\} = 0.65, \quad \text{prob}\{\xi_8(i,j) = 2\} = 0.35,$$
$$\text{prob}\{\xi_9(i,j) = 1\} = 0.35, \quad \text{prob}\{\xi_9(i,j) = 2\} = 0.65,$$
$$\text{prob}\{\xi_{10}(i,j) = 1\} = 0.75, \quad \text{prob}\{\xi_{10}(i,j) = 2\} = 0.25.$$

The noises $w(i,j)$ and $v_l(i,j)$ are uncorrelated Gaussian sequences with variances $Q(i,j) = 0.16$ and $R(i,j) = 0.09I$. The zero-mean random variables $\alpha_s(i,j)$ $(s = 1,2)$ and $\beta_l(i,j)$ $(l \in [1\ 10])$ are uncorrelated Gaussian sequences with $\sigma_{\alpha_s} = 0.04$ and $\sigma_{\beta_l} = 0.121$. The considered nonlinearities are given as $f(x(i,j)) = |x(i,j)|$ and $h_l(x(i,j)) = 0.25|x(i,j)|$, where $|x(i,j)|$ is a column vector whose elements are the absolute values of the corresponding elements in $x(i,j)$, that is, $|x(i,j)| = [|x_1(i,j)|\ |x_2(i,j)|]^T$. Then, the conditions (7.3a)–(7.3c) are satisfied with $\mu = 1$ and $\tau_l = 0.25$. Moreover, the initial conditions follow the Gaussian distribution with $u_1(i) = u_2(j) = [0\ 0]^T$ and $X(i,0) = X(0,j) = 0.04I$, and hence $P(i,0) = P(0,j) = 0.04I$. Besides, in this example, the scaling scalars are set as $\lambda = 1$ and $\xi = 0.01$.

According to Theorem 7.2, the distributed filter gains can be iteratively calculated as listed in Table 7.1 (only part of the results are presented here for space consideration). Simulation results are given in Figures 7.2–7.5. Particularly, the error trajectories of the first sensor node are depicted in Figures 7.2–7.3, where $e_1^{(1)}(i,j)$ and $e_1^{(2)}(i,j)$ are respectively the first and second elements of $e_1(i,j)$. Figure 7.4 shows the trace of the locally minimal upper bound $M(i,j)$ under the designed gain parameters. Figure 7.5 plots the trace of the estimation error variance $P(i,j)$ in the mean square sense from 1,000 Monte Carlo runs. The simulation results confirm that the proposed filter works well with a favorable filtering performance.

Obviously, the established bound $M(i,j)$ depends on the scaling parameter λ (that is commonly preset and adjustable for specific application requirements). By respectively selecting $\lambda = 0.5$ and $\lambda = 1.5$ without changing the

TABLE 7.1: Part of the filter gains

$K_{11}(1,2) =$	0.1853	0.2703	$K_{11}(2,1) =$	0.1982	0.2924
	0.1747	0.2527		0.1551	0.2286
$K_{22}(1,2) =$	−0.2112	0.2372	$K_{22}(2,1) =$	−0.2490	0.2788
	0.0563	0.0827		0.0496	0.0648
$K_{33}(1,2) =$	0.2997	−0.0463	$K_{33}(2,1) =$	0.3499	−0.0346
	0.0391	−0.0662		0.0294	−0.0538
$K_{44}(1,2) =$	0.3824	0.2800	$K_{44}(2,1) =$	0.4299	0.3295
	−0.0026	0.0019		−0.0039	−0.0030

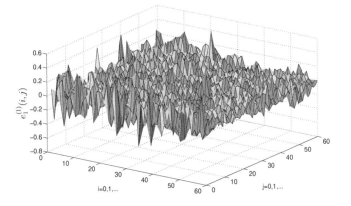

FIGURE 7.2: Estimation error $e_1^{(1)}(i,j)$ for sensor node 1.

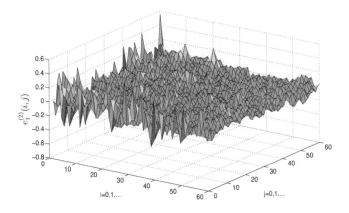

FIGURE 7.3: Estimation error $e_1^{(2)}(i,j)$ for sensor node 1.

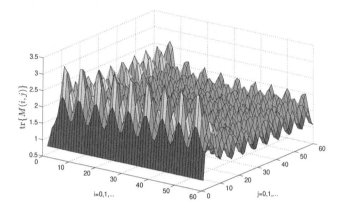

FIGURE 7.4: Trace trajectory of $M(i,j)$.

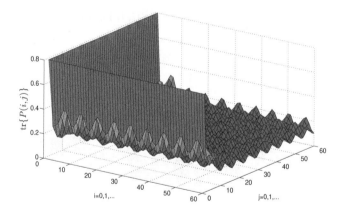

FIGURE 7.5: Trace trajectory of $P(i,j)$.

other parameters, the relevant simulation results are shown in Figs. 7.6–7.7. It can be seen that the scaling scalar λ affects the filtering performance.

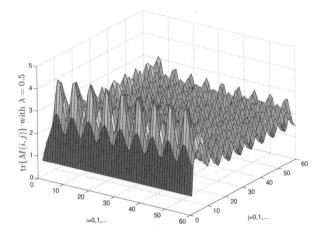

FIGURE 7.6: Trace trajectory of $M(i,j)$ with $\lambda = 0.5$.

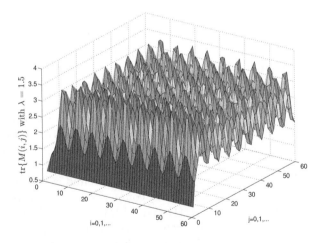

FIGURE 7.7: Trace trajectory of $M(i,j)$ with $\lambda = 1.5$.

For the ideal situation where the shared network media have no bandwidth limitation (i.e., it is unnecessary to adopt the RA protocol), the filtering performance is intuitively expected to be better than the counterpart in the practical case with the proposed RA protocol. Apparently, in such an ideal case, one has $\Lambda_{\xi_l(i,j)} = I$ and thus $\Pi_l = I$, $\hat{\Pi}_l = \mathbf{11^T}$ and $\tilde{\Pi} = 0$, under which the scalar ξ introduced in the constrained bandwidth situation (to ensure the positive definiteness of matrix $\hat{R}(i,j)$ for facilitating the filter design) is not required anymore. With purpose of evaluating the influence of the RA protocol programming, Fig. 7.8 is given confirming that the locally minimal upper bound $M(i,j)$ in the ideal case is indeed tighter as compared with that presented in Fig. 7.4.

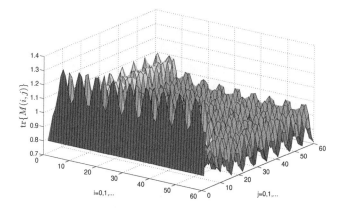

FIGURE 7.8: Trace trajectory of $M(i,j)$ without the RA protocol.

To evaluate the noise effect, we reset $R(i,j) = 0.49$ while remaining all the other parameters. In this case, the trace of the locally minimal upper bound $M(i,j)$ is presented in Fig. 7.9. Comparing with Fig. 7.4 and Fig. 7.9, we note that a greater value of $R(i,j)$ leads to a larger upper bound, namely, a worse filtering performance.

To further assess the distributed filtering strategy, two additional cases are considered. One is that the sensor network is completely connected, namely $\mathcal{A} = \mathbf{11^T}$, where the local minimization of upper bound on the error variance is denoted as $M_1(i,j)$. The other one is that there are no neighbors for all sensors, namely $\mathcal{A} = I$, where the locally optimal upper bound is denoted as $M_0(i,j)$. In the two scenarios, the corresponding comparisons are provided, where the trajectories of $\mathrm{tr}\{M(i,j) - M_1(i,j)\}$ and $\mathrm{tr}\{M_0(i,j) - M(i,j)\}$ are presented in Fig. 7.10 and Fig. 7.11, respectively. It is concluded that the performance of the proposed filtering scheme is closely linked to the network sparsity. More specifically, the designed filter performs best with $\mathcal{A} = \mathbf{11^T}$ whereas worst with $\mathcal{A} = I$ as taking the available and exchangeable information acquired by each sensor in different cases into account.

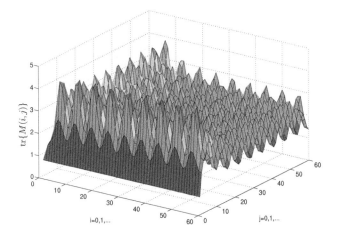

FIGURE 7.9: Trace trajectory of $M(i,j)$ with $R(i,j) = 0.49I$.

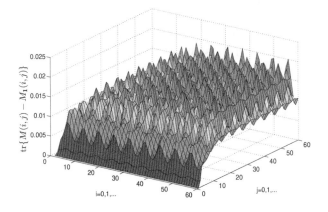

FIGURE 7.10: Difference between the traces of $M(i,j)$ and $M_1(i,j)$.

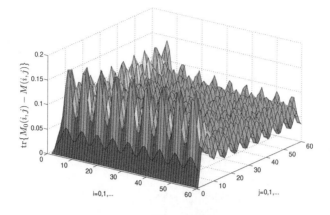

FIGURE 7.11: Difference between the traces of $M_0(i,j)$ and $M(i,j)$.

7.4 Summary

The distributed filtering problem has been investigated for 2-D systems with shift-varying parameters, stochastic nonlinearities, and RA protocol over a sensor network with given topology. The proposed protocol scheduling is modeled by a sequence of random variables which is mutually uncorrelated and obeys certain probability distributions. Note that, in the presence of stochastic nonlinearities, it is intractable to obtain the analytical expression of the filtering error variance, let alone the design of the distributed filters. As such, the tightest upper bound on the estimation error variance (in the trace sense) has been regarded as an alternative filtering performance. Based on the inductive method, certain upper bound of the estimation error variance has been first derived, and then the filter parameters have been designed to locally minimize the obtained upper bound by using the matrix simplification technique. An example of practical insight has also been presented to demonstrate the effectiveness of the proposed filter design scheme.

8

Resilient Filtering for Linear Shift-Varying Repetitive Processes under Uniform Quantizations and Round-Robin Protocols

The LRP as a distinct type of 2-D systems has recently drawn considerable research attention due to its broad applications in varieties of industrial systems. A typical LRP is dominated by a sequence of sweeps along with their process outputs through differential or difference dynamics defined over a finite duration. In technical terms, the sweep, the process output, and the duration are called the batch/pass, the pass profile, and the pass length, respectively. The process output is generated on each pass and works as a forcing function, which conduces to the dynamics of the next pass profile. The distinguishing feature of the LRP consists in two aspects: (1) its dynamics along passes broadcasts over a finite duration and (2) the pass profile exhibits from pass to pass, which results in the two-directional characteristic.

To date, the filtering problem concerning LRPs has received a steadily growing research interest and a few filter design strategies have been available in the literature [7, 19, 20, 226]. In particular, the H_∞ state estimation problem has been investigated in [226] for a class of continuous-time LRPs to guarantee a prescribed disturbance attenuation in the worst case. The moving-horizon estimation algorithm has been employed in [7] as an optimization-based method by using a fixed-size moving-window for estimation. The Kalman filtering has been considered in [20] with focus on obtaining the local minimization of the estimation error variance. The robust Kalman filter has been proposed in [19] to deal with the repetitive processes with uncertain parameters.

It is well-known that communication constraints have become a major reason leading to various undesired network-induced phenomena. A crucial issue for networked systems is to meliorate the utilization efficiency of the limited communication bandwidths over a resource-scarce network. With this purpose, the scheduling protocols have been widely adopted in industry to regulate the data transmission for saving network bandwidth. Among various scheduling protocols, the RR protocol is of practical importance since it predefines a periodic transmission rule which makes the scheduling easy-to-implement. In comparison to the conventional communication without protocol, consideration of the RR scheduling would inevitably lead to certain challenges in the analysis/synthesis of the filter performance. Up to now, some preliminary results have been available for the control/estimation problems with RR scheduling by means of periodic switching method or accumulated delay approach [114,127,265]. Nevertheless, when it comes to the recursive filtering for shift-varying LRPs, the corresponding results have been very scarce, not to mention the case where filter resilience and uniform quantization are also taken into simultaneous account.

In view of the discussions made so far, in this chapter, we attempt to address the resilient filtering problem for shift-varying LRPs with RR protocol and uniform quantization. The addressed problem appears to be challenging as we would have to answer the following four questions: (1) how to modify the traditional RR protocol to adapt the repetitive processes which are a special class of 2-D systems? (2) how to handle the quantization effect and the RR protocol involved in the complicated dynamic behaviors? (3) how to design a recursive filter to fulfill the desired filtering performance? and (4) how to evaluate the filtering performance of the proposed resilient filter? To this end, the aim of this chapter is to offer satisfactory responses to these questions.

In this chapter, the shift-varying LRP under consideration is transformed into a general FM-II model and an RR protocol law is defined by using an ordered time sequence for the LRP. The signal transmission from sensors to the remote filters suffers from uniform quantization through a shared network, where only one sensor is allowed to transmit information at each time step. According to intensive stochastic analysis and mathematical induction, sufficient criteria are derived under which a certain upper bound on the filtering error variance is developed and subsequently minimized, and the boundedness of the error variance is also analyzed. The main novelties of this chapter are stated as follows: (1) compared with the existing literature, the resilient filter design problem is, for the first time, investigated for the shift-varying LRP with uniform quantization and RR protocol scheduling; (2) the locally minimal upper bound is constructed for the estimation error variance and the filter design scheme is presented in a recursive form which is easy-to-implement; and (3) the boundedness of the filtering performance is elaborately discussed for the designed resilient filter.

The rest of this chapter is organized as follows. Section 8.1 formulates the resilient filter design problem for the LRPs subject to uniform quantization and RR protocol. The main results are presented in Section 8.2 where the desired filter is derived and the boundedness analysis of the filtering performance is given. Section 8.3 provides a numerical example, and Section 8.4 draws the conclusions.

8.1 Problem Formulation

8.1.1 The System Model

Consider the following shift-varying LRP defined on $l \in [0\ \infty)$ and $k \in [0\ \hbar]$ with $\hbar$ being the pass length:

$$\vec{x}(l+1, k+1) = \vec{A}_1(l+1, k)\vec{x}(l+1, k) + \vec{B}_1(l, k)\vec{y}(l, k)$$
$$+ \vec{C}_1(l+1, k)w(l+1, k), \tag{8.1a}$$
$$\vec{y}(l+1, k) = \vec{A}_2(l+1, k)\vec{x}(l+1, k) + \vec{B}_2(l, k)\vec{y}(l, k)$$
$$+ \vec{C}_2(l+1, k)w(l+1, k), \tag{8.1b}$$

where $\vec{x}(l+1, k) \in \mathbb{R}^n$ and $\vec{y}(l, k) \in \mathbb{R}^m$ are the state vector and the pass profile vector, respectively; $w(l+1, k) \in \mathbb{R}^p$ is the process noise with zero mean and variance $Q(l+1, k)$; and $\vec{A}_s(l+1, k)$, $\vec{B}_s(l, k)$, and $\vec{C}_s(l+1, k)$ ($s = 1, 2$) are known shift-varying matrices with appropriate dimensions.

For $l \in [0\ \infty)$ and $k \in [0\ \hbar]$, the measurement before transmittal via the network medium is described as follows:

$$z(l+1, k) = \vec{D}_1(l+1, k)\vec{x}(l+1, k) + \vec{D}_2(l, k)\vec{y}(l, k) + v(l+1, k) \tag{8.2}$$

where $z(l+1, k) \in \mathbb{R}^q$ is the measured output vector, $v(l+1, k) \in \mathbb{R}^q$ is the measurement noise with zero mean and variance $R(l+1, k)$, and $\vec{D}_s(l, k)$ ($s = 1, 2$) are known matrices.

The initial conditions concerning (8.1) are given as:

$$\mathbb{E}\{\vec{x}(l+1, 0)\} = d(l+1), \quad \mathbb{E}\{\vec{y}(0, k)\} = f(k) \tag{8.3}$$

where $d(l+1) \in \mathbb{R}^n$ and $f(k) \in \mathbb{R}^m$ are known vectors.

Remark 8.1 *The classical LRP is described by a series of passes whose dynamics evolves over a finite duration viewed as the pass length. The pass profile, namely the process output, is developed based on the state related to the current pass, which plays a role as a forcing function in the dynamics of the next pass profile. Notice that the state of LRP broadcasts along each pass over a fixed pass length, while the pass profile evolves along the pass-to-pass direction. Thanks to such a two-directional transmission feature, the LRP has been recognized as a special type of 2-D systems and then investigated via 2-D theory in some existing literature.*

For convenience of technical analysis, we convert systems (8.1)–(8.2) into a general FM-II model. Define

$$x(l, k) \triangleq [\vec{x}^T(l, k) \quad \vec{y}^T(l-1, k)]^T,$$

$$A_1(l,k) \triangleq \begin{bmatrix} \vec{A}_1(l,k) & \vec{B}_1(l-1,k) \\ 0 & 0 \end{bmatrix},$$

$$A_2(l,k) \triangleq \begin{bmatrix} 0 & 0 \\ \vec{A}_2(l,k) & \vec{B}_2(l-1,k) \end{bmatrix},$$

$$C_1(l,k) \triangleq \begin{bmatrix} \vec{C}_1(l,k) \\ 0 \end{bmatrix}, \quad C_2(l,k) \triangleq \begin{bmatrix} 0 \\ \vec{C}_2(l,k) \end{bmatrix},$$

$$D(l,k) \triangleq [\vec{D}_1(l,k) \;\; \vec{D}_2(l-1,k)].$$

Then, for $l \in [2 \; \infty)$ and $k \in [1 \; \hbar]$, it follows from (8.1) and (8.2) that

$$
\begin{aligned}
x(l,k) &= A_1(l,k-1)x(l,k-1) + A_2(l-1,k)x(l-1,k) \\
&\quad + C_1(l,k-1)w(l,k-1) + C_2(l-1,k)w(l-1,k), \quad\quad (8.4a)
\end{aligned}
$$
$$
z(l,k-1) = D(l,k-1)x(l,k-1) + v(l,k-1). \quad\quad (8.4b)
$$

The initial conditions related to (8.4) are set as

$$
x(l,0) = \begin{bmatrix} \vec{x}(l,0) \\ \vec{y}(l-1,0) \end{bmatrix}, \quad l \in [1 \; \infty) \quad\quad (8.5a)
$$

$$
x(1,k) = \begin{bmatrix} \vec{x}(1,k) \\ \vec{y}(0,k) \end{bmatrix}, \quad k \in [0 \; \hbar] \qu\quad (8.5b)
$$

and the corresponding statistics of $\mathbb{E}\{x(l,0)\}$ and $\mathbb{E}\{x(1,k)\}$ can be calculated from (8.1) and (8.3).

8.1.2 Network Description

The data transmissions inevitably suffer from certain network constraints over the communication medium of limited capacity. In this chapter, the uniform quantization phenomenon is considered as a reflection of the limited bandwidth. In addition, to prevent possible data collision and fully utilize the communication resource, the RR protocol is adopted to prescribe a transmission order for sensor nodes.

To begin with, an ordered time sequence shall be defined for system (8.4). Recalling the operation of LRP, for time instants (l_1, k_1) and (l_2, k_2), it is reasonable to define the following relationships:

$(l_1, k_1) < (l_2, k_2)$ if and only if

$(l_1, k_1) \in \{(l,k)|l = l_2, k < k_2\} \cup \{(l,k)|l < l_2, k \in [1 \; \hbar]\}$

and

$$(l_1, k_1) = (l_2, k_2) \text{ if and only if } l_1 = l_2 \text{ and } k_1 = k_2.$$

Based on the above descriptions, the chronological order is determined for any given time instants.

It is now ready to describe the measurements subject to the considered networked effects, where the signals broadcasted through the network medium are quantized first and then dispatched in a certain rule based on the RR protocol.

The influence of uniform quantization is presented as follows. Suppose that the quantization range is confined on the interval $[-\mathcal{S},\ \mathcal{S}]$ with $\mathcal{S} > 0$, and the length of the quantization level is set as $\tau = 2\mathcal{S}/2^{\eta}$, where η is the number of bits for sensors. Let us rewrite the measurement $z(l, k)$ as follows:

$$z(l,k) = [z_1(l,k)\ z_2(l,k)\ \cdots\ z_q(l,k)]^T,$$

where $z_s(l, k)$ $(s \in [1\ q])$ is the measurement received by the s-th sensor before being transmitted through the network medium. The quantized measurement is then denoted as:

$$\bar{\mathcal{Q}}(z(l,k)) = [\mathcal{Q}(z_1(l,k))\ \mathcal{Q}(z_2(l,k))\ \cdots\ \mathcal{Q}(z_q(l,k))]^T$$

where

$$\mathcal{Q}(z_s(l,k)) \triangleq \tau\Re(z_s(l,k)/\tau), \quad s \in [1\ q]$$

in which $\Re(\cdot)$ is a stochastic function of probabilistic uniform type. As discussed in [95, 233], when $i\tau \leq z_s(l,k) \leq (i+1)\tau$ with i belonging to $[-2^{\eta-1}\ 2^{\eta-1}]$, the s-th measurement $z_s(l, k)$ is quantized to satisfy

$$\text{prob}\{\mathcal{Q}(z_s(l,k)) = i\tau\} = 1 - \rho_s,$$
$$\text{prob}\{\mathcal{Q}(z_s(l,k)) = (i+1)\tau\} = \rho_s$$

with $\rho_s = (z_s(l,k) - i\tau)/\tau$. Define the quantization error as

$$\begin{aligned}
e_z(l,k) &\triangleq \bar{\mathcal{Q}}(z(l,k)) - z(l,k) \\
&\triangleq [e_{z1}(l,k)\ e_{z2}(l,k)\ \cdots\ e_{zq}(l,k)]^T.
\end{aligned} \tag{8.6}$$

Then, for $s \in [1\ q]$, $e_{zs}(l, k)$ can be modeled by the following Bernoulli distribution:

$$\text{prob}\{e_z(l,k) = -\rho_s\tau\} = 1 - \rho_s, \tag{8.7}$$
$$\text{prob}\{e_z(l,k) = (1 - \rho_s)\tau\} = \rho_s. \tag{8.8}$$

It is worth mentioning that the errors $e_{zs}(l, k)$ are recognized as q independent random sequences since the quantization phenomenon is encountered locally for each component. According to (8.7)–(8.8), one has $\mathbb{E}\{e_{zs}(l,k)\} = 0$ and $\text{Var}\{e_{zs}(l,k)\} \leq \frac{1}{4}\tau^2$ for all $s \in [1\ q]$.

Next, taking the communication protocol into account, we introduce the RR scheduling for signal regulations in order to mitigate/avoid data collisions. Recall that the main idea of the RR scheduling is to arrange the transmission order of sensors one by one in a circle sequence. The application of RR protocol to repetitive processes can be executed by means of the predefined chronological order. Particularly, let $\xi(l,k) \in [1 \ q]$ be the selected sensor getting access to the network medium at time instant (l,k). Based on the RR protocol, the selection of $\xi(l,k)$ is chosen as

$$\xi(l,k) = \text{mod}(l\hbar + k - 1, q) + 1 \qquad (8.9)$$

which infers that the s-th sensor is allowed to send information if and only if $\text{mod}(l\hbar + k - s, q) = 0$. The measurement after transmission is denoted as

$$\bar{z}(l,k) \triangleq [\bar{z}_1(l,k) \ \bar{z}_2(l,k) \cdots \bar{z}_q(l,k)]^T$$

where the update of $\bar{z}_s(l,k)$ $(s \in [1 \ q])$ obeys

$$\bar{z}_s(l,k) = \begin{cases} \mathcal{Q}(z_s(l,k)), & \text{if } \text{mod}(l\hbar + k - s, q) = 0 \\ \bar{z}_s(l,k-1), & \text{otherwise} \end{cases} \qquad (8.10)$$

with $\bar{z}_s(l,k) = 0$ for any $k < 0$. From (8.9) to (8.10), the expression of $\bar{z}(l,k)$ is presented as

$$\bar{z}(l,k) = \Phi_{\xi(l,k)}\bar{\mathcal{Q}}(z(l,k)) + (I - \Phi_{\xi(l,k)})\bar{z}(l,k-1), \qquad (8.11)$$

where

$$\Phi_{\xi(l,k)} \triangleq \text{diag}\{\delta(\xi(l,k),1), \delta(\xi(l,k),2), \ldots, \delta(\xi(l,k),q)\}$$

and $\delta(\cdot,\cdot)$ is the Kronecker delta function.

Remark 8.2 *The actually measured output is established in (8.11) under the network-induced effects of uniform quantization and RR protocol. Broadly speaking, there are two strategies to tackle the measurements after transmission through the shared medium, namely, the zero input strategy and the zero-order hold (ZOH) like strategy. Here, as a popular compensation mechanism in practice, the ZOH like strategy is utilized to offset the measurements that do not acquire the network access at the current transmission time. It can be observed from (8.11) that $\bar{z}(l,k)$ consists of two terms, where $\bar{\mathcal{Q}}(z(l,k))$ in the first term is produced by the quantization effect, $\bar{z}(l,k-1)$ involved in the second term results from the ZOH like strategy, and $\Phi_{\xi(l,k)}$ reflects the token-dependent scheduling. It should be noted that the quantization phenomenon and the ZOH like strategy will inevitably lead to substantial difficulties in analyzing the system dynamics and the filtering performance.*

For simplicity, a compact form of the addressed system is given for later developments. By denoting $\bar{x}(l,k) \triangleq [x^T(l,k) \ \bar{z}^T(l,k-1)]^T$ and $\bar{w}(l,k) \triangleq [w^T(l,k) \ v^T(l,k)]^T$, one obtains from (8.4a), (8.6), and (8.11) that

$$\bar{x}(l,k) = \bar{A}_1(l,k-1)\bar{x}(l,k-1) + \bar{A}_2(l-1,k)\bar{x}(l-1,k)$$

$$+ \bar{C}_1(l, k-1)\bar{w}(l, k-1) + \bar{C}_2(l-1, k)$$
$$\times \bar{w}(l-1, k) + \bar{\Phi}_{\xi(l,k-1)}e_z(l, k-1), \tag{8.12a}$$
$$\bar{z}(l, k) = \Gamma_{\xi(l,k)}\bar{x}(l, k) + \Phi_{\xi(l,k)}(v(l, k) + e_z(l, k)), \tag{8.12b}$$

where

$$\bar{A}_1(l, k) \triangleq \begin{bmatrix} A_1(l, k) & 0 \\ \Phi_{\xi(l,k)}D(l, k) & I - \Phi_{\xi(l,k)} \end{bmatrix},$$
$$\bar{A}_2(l, k) \triangleq \mathrm{diag}\{A_2(l, k), 0\},$$
$$\bar{C}_1(l, k) \triangleq \mathrm{diag}\{C_1(l, k), \Phi_{\xi(l,k)}\},$$
$$\bar{C}_2(l, k) \triangleq \mathrm{diag}\{C_2(l, k), 0\}, \quad \bar{\Phi}_{\xi(l,k)} \triangleq \begin{bmatrix} 0 \\ \Phi_{\xi(l,k)} \end{bmatrix},$$
$$\Gamma_{\xi(l,k)} \triangleq \begin{bmatrix} \Phi_{\xi(l,k)}D(l, k) & I - \Phi_{\xi(l,k)} \end{bmatrix}.$$

8.1.3 Resilient Filter

This chapter is concerned with the recursive filtering problem for the LRP with network-induced effects. The system diagram is shown in Fig. 8.1, where the measurements are subjected to quantization firstly and scheduled subsequently by the RR protocol over a shared communication medium.

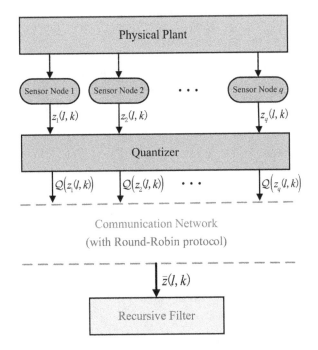

FIGURE 8.1: Block diagram for the LRP over the communication network.

A Kalman-type filter is adopted for (8.12) of the following structure with $l \in [2 \; \infty)$ and $k \in [1 \; \hbar]$:

$$\hat{x}_p(l, k) = \bar{A}_1(l, k-1)\hat{x}_u(l, k-1) + \bar{A}_2(l-1, k)\hat{x}_u(l-1, k), \qquad (8.13a)$$

$$\hat{x}_u(l, k) = \hat{x}_p(l, k) + (K(l, k) + \Delta(l, k))\left(\bar{z}(l, k) - \Gamma_{\xi(l,k)}\hat{x}_p(l, k)\right), \qquad (8.13b)$$

where $\hat{x}_p(l, k)$ and $\hat{x}_u(l, k)$ are referred to as the prediction and the estimate of $\bar{x}(l, k)$, respectively; $K(l, k)$ is the filter gain to be designed; and $\Delta(l, k)$ represents the possible gain perturbation. Inspired by [76], the gain perturbation here is modeled by

$$\Delta(l, k) = \sum_{i=1}^{r} \theta_i(l, k) K_i,$$

where K_i ($i \in [1 \; r]$) are known matrices and $\theta_i(l, k)$ are r mutually uncorrelated random scalar sequences with $\mathbb{E}\{\theta_i(l, k)\} = 0$ and $\mathrm{Var}\{\theta_i(l, k)\} = \sigma_{\theta_i}$. The initial conditions for (8.13) are given as $\mathbb{E}\{\hat{x}_u(l, 0)\} = \mathbb{E}\{\hat{x}_u(1, k)\} = 0$ for $l \in [1 \; \infty)$ and $k \in [0 \; \hbar]$.

Remark 8.3 *Since the computational errors or implementation uncertainties might be encountered during the hardware realization of the gain parameter, it would be impossible to obtain the exact filter gain. Hence, the filter to be designed shall have resilience against certain perturbation. As such, the resilient filtering design problem is considered, where the matrix $\Delta(l, k)$ is introduced to model the gain perturbation.*

Assumption 8.1 *For all $l_\zeta \in [0 \; \infty)$, $k_\zeta \in [0 \; \hbar]$ ($\zeta \in [0 \; 5]$), $s \in [1 \; q]$, and $i \in [1 \; r]$, the random variables $\bar{x}(l_1, 0)$, $\bar{y}(0, k_1)$, $w(l_2, k_2)$, $v(l_3, k_3)$, $e_{zs}(l_4, k_4)$, and $\theta_i(l_5, k_5)$ are white noise sequences, which are mutually uncorrelated with each other.*

Define

$$e_p(l, k) \triangleq x(l, k) - \hat{x}_p(l, k),$$

$$e_u(l, k) \triangleq x(l, k) - \hat{x}_u(l, k),$$

$$P_p(l, k) \triangleq \mathbb{E}\left\{e_p(l, k)e_p^T(l, k)\right\},$$

$$P_u(l, k) \triangleq \mathbb{E}\left\{e_u(l, k)e_u^T(l, k)\right\}.$$

It follows from (8.13) that

$$
\begin{aligned}
e_p(l, k) = &\bar{A}_1(l, k-1)e_u(l, k-1) + \bar{A}_2(l-1, k)e_u(l-1, k) \\
&+ \bar{C}_1(l, k-1)\bar{w}(l, k-1) + \bar{C}_2(l-1, k) \\
&\times \bar{w}(l-1, k) + \bar{\Phi}_{\xi(l,k-1)}e_z(l, k-1), \qquad (8.14a) \\
e_u(l, k) = &\left[I - (K(l, k) + \Delta(l, k))\Gamma_{\xi(l,k)}\right] e_p(l, k)
\end{aligned}
$$

$$- (K(l,k) + \Delta(l,k))\Phi_{\xi(l,k)}(v(l,k) + e_z(l,k)). \tag{8.14b}$$

Note that the existence of the quantization error results in the inaccessibility of the analytical solution to the error variance $P_u(l,k)$. Therefore, the main objective of the addressed problem is to design the resilient filter (8.13) such that certain upper bound of $P_u(l,k)$ is first ensured and then minimized at each instant.

8.2 Main Results

In this section, the resilient filtering problem is investigated for the LRP with uniform quantization and RR scheduling. First, the evolutions of the error variances $P_p(l,k)$ and $P_u(l,k)$ are presented. Then, with the aid of two Riccati-like equations, an upper bound on the filtering error variance $P_u(l,k)$ is established which is locally optimized by properly designing the filter gain. Moreover, the boundedness of the error variance is discussed with rigorous analysis.

The following lemma is provided for the subsequent developments.

Lemma 8.1 *For the error dynamics (8.14), the recursions of the error variances $P_p(l,k)$ and $P_u(l,k)$ are given as follows:*

$$\begin{aligned}
P_p(l,k) =\; & \bar{A}_1(l,k-1)P_u(l,k-1)\bar{A}_1^T(l,k-1) \\
& + \bar{A}_2(l-1,k)P_u(l-1,k)\bar{A}_2^T(l-1,k) \\
& + \bar{A}_1(l,k-1)\mathbb{E}\left\{e_u(l,k-1)e_u^T(l-1,k)\right\}\bar{A}_2^T(l-1,k) \\
& + \bar{A}_2(l-1,k)\mathbb{E}\left\{e_u(l-1,k)e_u^T(l,k-1)\right\}\bar{A}_1^T(l,k-1) \\
& + \bar{\Phi}_{\xi(l,k-1)}Z(l,k-1)\bar{\Phi}_{\xi(l,k-1)}^T \\
& - \bar{A}_1(l,k-1)K(l,k-1)\Phi_{\xi(l,k-1)}Z(l,k-1)\bar{\Phi}_{\xi(l,k-1)}^T \\
& - \bar{\Phi}_{\xi(l,k-1)}Z(l,k-1)\Phi_{\xi(l,k-1)}^T K^T(l,k-1)\bar{A}_1^T(l,k-1) \\
& - \bar{A}_1(l,k-1)\hat{R}(l,k-1)\bar{C}_1^T(l,k-1) \\
& - \bar{C}_1(l,k-1)\hat{R}^T(l,k-1)\bar{A}_1^T(l,k-1) \\
& - \bar{A}_2(l-1,k)\hat{R}(l-1,k)\bar{C}_2^T(l-1,k) \\
& - \bar{C}_2(l-1,k)\hat{R}^T(l-1,k)\bar{A}_2^T(l-1,k) \\
& + \bar{C}_1(l,k-1)\bar{Q}(l,k-1)\bar{C}_1^T(l,k-1) \\
& + \bar{C}_2(l-1,k)\bar{Q}(l-1,k)\bar{C}_2^T(l-1,k), \tag{8.15}
\end{aligned}$$

$$\begin{aligned}
P_u(l,k) =\; & \left(I - K(l,k)\Gamma_{\xi(l,k)}\right)P_p(l,k)\left(I - K(l,k)\Gamma_{\xi(l,k)}\right)^T \\
& + \sum_{i=1}^{r}\sigma_{\theta_i}K_i\left[\Gamma_{\xi(l,k)}P_p(l,k)\Gamma_{\xi(l,k)}^T + \bar{R}(l,k)\right]K_i^T
\end{aligned}$$

$$+ K(l,k)\bar{R}(l,k)K^T(l,k), \qquad (8.16)$$

where

$$Z(l,k) \triangleq \mathbb{E}\left\{e_z(l,k)e_z^T(l,k)\right\},$$
$$\bar{Q}(l,k) \triangleq \mathrm{diag}\{Q(l,k), R(l,k)\},$$
$$\hat{R}(l,k) \triangleq K(l,k)\Phi_{\xi(l,k)}[0_{q\times p} \ R(l,k)],$$
$$\bar{R}(l,k) \triangleq \Phi_{\xi(l,k)}(R(l,k) + Z(l,k))\Phi_{\xi(l,k)}^T.$$

Proof *Recall that the components of the quantization error $e_z(l,k)$ are mutually independent and the uncorrelatedness holds among $e_z(l,k)$, $\bar{w}(l,k)$, and $\Delta(l,k)$ from Assumption 8.1. One knows that the quantization error $e_z(l,k)$ is uncorrelated with the one-step prediction error $e_p(l,k)$. Furthermore, it is easy to derive from (8.14b) and the statistical properties*

$$\mathbb{E}\{e_z(l,k)\} = 0, \quad \mathbb{E}\{\bar{w}(l,k)\} = 0, \quad \mathbb{E}\{\Delta(l,k)\} = 0,$$
$$\mathbb{E}\{\bar{w}(l,k)\bar{w}^T(l,k)\} = \bar{Q}(l,k)$$

that

$$\mathbb{E}\left\{e_u(l,k-1)e_z^T(l,k-1)\right\}$$
$$= -K(l,k-1)\Phi_{\xi(l,k-1)}\mathbb{E}\left\{e_z(l,k-1)e_z^T(l,k-1)\right\}$$
$$= -K(l,k-1)\Phi_{\xi(l,k-1)}Z(l,k-1).$$

Then, according to the definition of $P_p(l,k)$, the validity of (8.15) is confirmed. In addition, it follows from (8.14b) that

$$P_u(l,k) = \mathbb{E}\left\{\left[I - (K(l,k) + \Delta(l,k))\Gamma_{\xi(l,k)}\right] P_p(l,k)\right.$$
$$\times \left[I - (K(l,k) + \Delta(l,k))\Gamma_{\xi(l,k)}\right]^T$$
$$\left. + (K(l,k) + \Delta(l,k))\bar{R}(l,k)(K(l,k) + \Delta(l,k))^T\right\}$$
$$= \left(I - K(l,k)\Gamma_{\xi(l,k)}\right) P_p(l,k) \left(I - K(l,k)\Gamma_{\xi(l,k)}\right)^T$$
$$+ \sum_{i=1}^{r} \sigma_{\theta_i} K_i \left[\Gamma_{\xi(l,k)} P_p(l,k)\Gamma_{\xi(l,k)}^T + \bar{R}(l,k)\right] K_i^T$$
$$+ K(l,k)\bar{R}(l,k)K^T(l,k)$$

which completes the proof.

Lemma 8.2 *[154] For matrices E, F, X, and H with appropriate dimensions, the following equalities hold:*

$$\frac{\partial \mathrm{tr}\{EXF\}}{\partial X} = E^T F^T, \qquad \frac{\partial \mathrm{tr}\{EX^T F\}}{\partial X} = FE,$$
$$\frac{\partial \mathrm{tr}\{EXFX^T H\}}{\partial X} = E^T H^T X F^T + HEXF.$$

8.2.1 The Upper Bounds and Filter Design

The following theorem is provided to construct certain upper bounds for the error variances and design the resilient filter gain that locally minimizes the obtained upper bounds.

Theorem 8.1 *Let α, ϵ, ϵ_1, ϵ_2, and μ be given positive scalars. Assume that there are two families of matrices $M_p(l,k) > 0$ and $M_u(l,k) > 0$ such that the following recursive equations*

$$
\begin{aligned}
M_p(l,k) &= (1+\alpha)\bar{A}_1(l,k-1)M_u(l,k-1)\bar{A}_1^T(l,k-1) \\
&\quad + (1+\alpha^{-1})\bar{A}_2(l-1,k)M_u(l-1,k)\bar{A}_2^T(l-1,k) \\
&\quad + \bar{C}_1(l,k-1)\left(\bar{Q}(l,k-1)+\epsilon_1\acute{R}(l,k-1)\right)\bar{C}_1^T(l,k-1) \\
&\quad + \bar{C}_2(l-1,k)\left(\bar{Q}(l-1,k)+\epsilon_2\acute{R}(l-1,k)\right)\bar{C}_2^T(l-1,k) \\
&\quad + \left(\epsilon_1^{-1}+\epsilon^{-1}(\tau^2/4)\right)\bar{A}_1(l,k-1)K(l,k-1)\Phi_{\xi(l,k-1)} \\
&\quad \times \Phi_{\xi(l,k-1)}^T K^T(l,k-1)\bar{A}_1^T(l,k-1) + (1+\epsilon)(\tau^2/4) \\
&\quad \times \bar{\Phi}_{\xi(l,k-1)}\bar{\Phi}_{\xi(l,k-1)}^T + \epsilon_2^{-1}\bar{A}_2(l-1,k)K(l-1,k) \\
&\quad \times \Phi_{\xi(l-1,k)}\Phi_{\xi(l-1,k)}^T K^T(l-1,k)\bar{A}_2^T(l-1,k), \quad (8.17) \\
M_u(l,k) &= \left(I - K(l,k)\Gamma_{\xi(l,k)}\right)M_p(l,k)\left(I-K(l,k)\Gamma_{\xi(l,k)}\right)^T \\
&\quad + \sum_{i=1}^{r}\sigma_{\theta_i}K_i\left[\Gamma_{\xi(l,k)}M_p(l,k)\Gamma_{\xi(l,k)}^T+\check{R}(l,k)\right]K_i^T \\
&\quad + K(l,k)(\check{R}(l,k)+\mu I)K^T(l,k) \quad (8.18)
\end{aligned}
$$

with initial constraints

$$
M_u(l,0) = P_u(l,0), \quad M_u(1,k) = P_u(1,k)
$$

are satisfied, where

$$
\acute{R}(l,k) \triangleq [0_{q\times p} \; R(l,k)]^T[0_{q\times p} \; R(l,k)],
$$

$$
\check{R}(l,k) \triangleq \Phi_{\xi(l,k)}\left(R(l,k)+\frac{\tau^2}{4}I\right)\Phi_{\xi(l,k)}^T.
$$

Then, for $l \in [2\ \infty)$ and $k \in [1\ \hbar]$, the error variances $P_p(l,k)$ and $P_u(l,k)$ are bounded by

$$
P_p(l,k) \le M_p(l,k), \quad P_u(l,k) \le M_u(l,k). \quad (8.19)
$$

Moreover, the local minimization of $M_u(l,k)$ is given by

$$
\begin{aligned}
M_u(l,k) &= M_p(l,k) - M_p(l,k)\Gamma_{\xi(l,k)}^T\tilde{R}^{-1}(l,k)\Gamma_{\xi(l,k)}M_p(l,k) \\
&\quad + \sum_{i=1}^{r}\sigma_{\theta_i}K_i\left[\Gamma_{\xi(l,k)}M_p(l,k)\Gamma_{\xi(l,k)}^T+\check{R}(l,k)\right]K_i^T \quad (8.20)
\end{aligned}
$$

with the gain parameter designed as

$$K(l, k) = M_p(l, k)\Gamma_{\xi(l,k)}^T \tilde{R}^{-1}(l, k) \tag{8.21}$$

in which $\tilde{R}(l, k) \triangleq \Gamma_{\xi(l,k)} M_p(l, k)\Gamma_{\xi(l,k)}^T + \check{R}(l, k) + \mu I$.

Proof *Based on (8.15) and $Z(l, k) \le (\tau^2/4)I$, one has*

$$\begin{aligned}
P_p(l, k) \le{} & (1 + \alpha)\bar{A}_1(l, k - 1)P_u(l, k - 1)\bar{A}_1^T(l, k - 1) \\
& + (1 + \alpha^{-1})\bar{A}_2(l - 1, k)P_u(l - 1, k)\bar{A}_2^T(l - 1, k) \\
& + \bar{C}_1(l, k - 1)(\bar{Q}(l, k - 1) + \epsilon_1\acute{R}(l, k - 1))\bar{C}_1^T(l, k - 1) \\
& + \bar{C}_2(l - 1, k)(\bar{Q}(l - 1, k) + \epsilon_2\acute{R}(l - 1, k))\bar{C}_2^T(l - 1, k) \\
& + \left(\epsilon_1^{-1} + \epsilon^{-1}(\tau^2/4)\right)\bar{A}_1(l, k - 1)K(l, k - 1)\Phi_{\xi(l,k-1)} \\
& \times \Phi_{\xi(l,k-1)}^T K^T(l, k - 1)\bar{A}_1^T(l, k - 1) + (1 + \epsilon)(\tau^2/4) \\
& \times \bar{\Phi}_{\xi(l,k-1)}\bar{\Phi}_{\xi(l,k-1)}^T + \epsilon_2^{-1}\bar{A}_2(l - 1, k)K(l - 1, k) \\
& \times \Phi_{\xi(l-1,k)}\Phi_{\xi(l-1,k)}^T K^T(l - 1, k)\bar{A}_2^T(l - 1, k). \tag{8.22}
\end{aligned}$$

The relationship $Z(l, k) \le (\tau^2/4)I$ also infers that $\bar{R}(l, k) \le \check{R}(l, k)$. Then, according to (8.17) and (8.22), we arrive at

$$\begin{aligned}
& P_p(l, k) - M_p(l, k) \\
& \le (1 + \alpha)\bar{A}_1(l, k - 1)(P_u(l, k - 1) - M_u(l, k - 1))\bar{A}_1^T(l, k - 1) \\
& \quad + (1 + \alpha^{-1})\bar{A}_2(l - 1, k)(P_u(l - 1, k) - M_u(l - 1, k))\bar{A}_2^T(l - 1, k) \tag{8.23}
\end{aligned}$$

and it follows from (8.16) and (8.18) that

$$\begin{aligned}
& P_u(l, k) - M_u(l, k) \\
& \le \left(I - K(l, k)\Gamma_{\xi(l,k)}\right)(P_p(l, k) - M_p(l, k))\left(I - K(l, k)\Gamma_{\xi(l,k)}\right)^T \\
& \quad + \sum_{i=1}^{r}\sigma_{\theta_i}K_i\Gamma_{\xi(l,k)}(P_p(l, k) - M_p(l, k))\Gamma_{\xi(l,k)}^T K_i^T. \tag{8.24}
\end{aligned}$$

In view of (8.23) and (8.24), it is evident to observe that, if $P_u(l, k - 1) \le M_u(l, k-1)$ and $P_u(l-1, k) \le M_u(l-1, k)$ are satisfied, then $P_p(l, k) \le M_p(l, k)$ is valid, which further infers that $P_u(l, k) \le M_u(l, k)$.

The assertion of (8.19) can be readily proven by applying the induction method pass-by-pass. According to the initial constraints $P_u(2, 0) \le M_u(2, 0)$ and $P_u(1, 1) \le M_u(1, 1)$, one has $P_p(2, 1) \le M_p(2, 1)$ and thus obtains $P_u(2, 1) \le M_u(2, 1)$. Assuming that $P_u(2, k_0) \le M_u(2, k_0)$ is true for a given integer $k_0 \in [1 \ \hbar - 1]$, we obtain from the initial condition $P_u(1, k_0 + 1) \le M_u(1, k_0 + 1)$ and (8.23) that

$$P_p(2, k_0 + 1) - M_p(2, k_0 + 1) \le 0$$

which further results in

$$P_u(2, k_0 + 1) - M_u(2, k_0 + 1) \leq 0.$$

Therefore, one has $P_u(2, k) \leq M_u(2, k)$ for $k \in [1 \; \hbar]$ from the induction.

Next, assume that $P_u(l, k) \leq M_u(l, k)$ is valid for $(l, k) \leq (l_0, \hbar)$ with a given $l_0 \in [2 \; \infty)$. It is easy to derive from $P_u(l_0 + 1, 0) \leq M_u(l_0 + l, 0)$ and $P_u(l_0, 1) \leq M_u(l_0, 1)$ that

$$P_p(l_0 + 1, 1) \leq M_p(l_0 + l, 1)$$

which also gives rise to the validity of

$$P_u(l_0 + 1, 1) \leq M_u(l_0 + l, 1).$$

In addition, with the assumption that $P_u(l_0 + 1, k_0) \leq M_u(l_0 + 1, k_0)$ for a given $k_0 \in [1 \; \hbar - 1]$, we have

$$
\begin{aligned}
&P_p(l_0 + 1, k_0 + 1) - M_p(l_0 + 1, k_0 + 1) \\
&\leq (1 + \alpha)\bar{A}_1(l_0 + 1, k_0)(P_u(l_0 + 1, k_0) - M_u(l_0 + 1, k_0)) \\
&\quad \times \bar{A}_1^T(l_0 + 1, k_0) + (1 + \alpha^{-1})\bar{A}_2(l_0, k_0 + 1) \\
&\quad \times (P_u(l_0, k_0 + 1) - M_u(l_0, k_0 + 1))\bar{A}_2^T(l_0, k_0 + 1) \\
&\leq 0
\end{aligned}
$$

which further contributes to

$$
\begin{aligned}
&P_u(l_0 + 1, k_0 + 1) - M_u(l_0 + 1, k_0 + 1) \\
&\leq \left(I - K(l_0 + 1, k_0 + 1)\Gamma_{\xi(l_0+1,k_0+1)}\right) \\
&\quad \times (P_p(l_0 + 1, k_0 + 1) - M_p(l_0 + 1, k_0 + 1)) \\
&\quad \times \left(I - K(l_0 + 1, k_0 + 1)\Gamma_{\xi(l,k)}\right)^T + \sum_{i=1}^{r} \sigma_{\theta_i} K_i \\
&\quad \times \Gamma_{\xi(l_0+1,k_0+1)}(P_p(l_0 + 1, k_0 + 1) - M_p(l_0 + 1, k_0 + 1)) \\
&\quad \times \Gamma_{\xi(l_0+1,k_0+1)}^T K_i^T \\
&\leq 0.
\end{aligned}
$$

Consequently, it follows from the induction that (8.19) is valid for all $l \in [2 \; \infty)$ and $k \in [1 \; \hbar]$.

To complete the proof of this theorem, the design of the filter gain that ensures the tightest upper bound remains to be determined. Taking the first variation to the trace of $M_u(l, k)$ regarding the gain parameter $K(l, k)$, we derive from Lemma 8.2 that

$$\frac{\partial \mathrm{tr}\{M_u(l, k)\}}{\partial K(l, k)} = 2K(l, k)\left[\Gamma_{\xi(l,k)} M_p(l, k)\Gamma_{\xi(l,k)}^T\right.$$

$$+ \check{R}(l,k) + \mu I] - 2M_p(l,k)\Gamma^T_{\xi(l,k)}. \qquad (8.25)$$

Recalling the semi-positive definite quadratic structure of $\mathrm{tr}\{M_u(l,k)\}$ *concerning* $K(l,k)$, *we set the right-hand side of (25) to be zero so as to minimize* $\mathrm{tr}\{M_u(l,k)\}$. *In this case, we obtain*

$$K(l,k) = M_p(l,k)\Gamma^T_{\xi(l,k)}\tilde{R}^{-1}(l,k)$$

which is consistent with (8.21). The proof is complete.

By virtue of the recursions (8.13), (8.17), (8.20)–(8.21) acquired so far, an overall algorithm can be outlined for designing the resilient filter with certain initial conditions. The flowchart of the resilient filtering algorithm is shown as in Fig. 8.2.

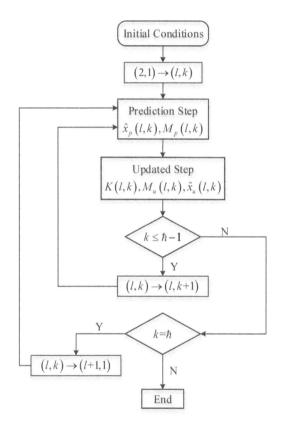

FIGURE 8.2: Flowchart of the proposed resilient filtering algorithm.

Remark 8.4 *Note that it is impossible to acquire the actual filtering error variance because of the quantization errors. In order to estimate the system state with an acceptable filtering performance, the locally tightest upper bound* of the filtering error variance is developed in Theorem 8.1 at each time instant by designing appropriate gain parameters. Using mathematical induction, the desired upper bound and filter gain are obtained by solving two sets of Riccati-like equations suitable for online calculation.

Remark 8.5 *The gain perturbation and the RR scheduling affect the locally minimal upper bound $M_u(l, k)$. More specifically, it can be seen from (8.18) that $M_u(l, k)$ is likely to get larger with the increase of σ_{θ_i} representing the statistical information of the gain perturbation. Additionally, under the proposed RR scheduling, a positive constant μ is introduced to ascertain the invertibility of matrix $\breve{R}(l, k)$. For the ideal case where all sensor nodes have access to the network medium at each transmission time, it is unnecessary to introduce the RR scheduling, and thus there is $\Phi_{\xi(l,k)} = I$. Therefore, one has $\breve{R}(l, k) = R(l, k) + (\tau^2/4)I > 0$, which further contributes to $\tilde{R}(l, k) = \Gamma_{\xi(l,k)} M_p(l, k) \Gamma_{\xi(l,k)}^T + \breve{R}(l, k) + \mu I > 0$ with $\mu = 0$. It follows from (8.20) that the decrease of $\tilde{R}(l, k)$ may help tighten the upper bound. Intuitively, a lower upper bound is expected in the ideal case, whereas $M_u(l, k)$ might be larger with less available measurements in a more realistic situation under the RR scheduling.*

8.2.2 Boundedness Analysis

After establishing the locally minimal upper bound for the error variance $P_u(l, k)$ at each instant, it would be interesting to evaluate the filter performance. In this subsection, the boundedness analysis is investigated with respect to $P_u(l, k)$. At the beginning, the following assumptions are made.

Assumption 8.2 *For $l \in [2 \ \infty)$ and $k \in [1 \ \hbar]$, matrix $D(l, k)$ has a full row rank.*

Assumption 8.3 *There are nonnegative real constants $\bar{a}_s$, $\bar{c}_s$ ($s = 1, 2$), $\bar{q}$, $\underline{d}$, $\bar{d}$, $\bar{r}$, and $\bar{k}$ such that the following relationships hold for $l \in [2 \ \infty)$, $k \in [1 \ \hbar]$, and $i \in [1 \ r]$:*

$$\bar{A}_s(l, k)\bar{A}_s^T(l, k) \leq \bar{a}_s I, \quad R(l, k) \leq \bar{r}I,$$
$$\bar{C}_s(l, k)\bar{C}_s^T(l, k) \leq \bar{c}_s I, \quad \bar{Q}(l, k) \leq \bar{q}I,$$
$$0 < \underline{d}I \leq D(l, k)D^T(l, k) \leq \bar{d}I, \quad K_i K_i^T \leq \bar{k}I.$$

Remark 8.6 *Here, Assumption 8.2 is a standard constraint that is valid in most engineering applications. Such an assumption results in the fact that $D(l, k)D^T(l, k) > 0$, and hence a positive scalar $\underline{d}$ can be found to ensure that $D(l, k)D^T(l, k) \geq \underline{d}I > 0$.*

Lemma 8.3 *Under Assumptions 8.2–8.3, the following relationship*

$$0 < \underline{\lambda}I \leq \Gamma_{\xi(l,k)}\Gamma_{\xi(l,k)}^T \leq \bar{\lambda}I \tag{8.26}$$

is satisfied for $l \in [2 \; \infty)$ and $k \in [1 \; \hbar]$, where

$$\underline{\lambda} \triangleq \min\{\underline{d}, 1\}, \quad \bar{\lambda} \triangleq \max\{\bar{d}, 1\}.$$

Proof *Recalling the expressions of $\Gamma_{\xi(l,k)}$ and $\Phi_{\xi(l,k)}$, under the RR scheduling, one easily derives that*

$$\begin{aligned}
\Gamma_{\xi(l,k)}\Gamma_{\xi(l,k)}^T &= \Phi_{\xi(l,k)}D(l,k)D^T(l,k)\Phi_{\xi(l,k)}^T + (I - \Phi_{\xi(l,k)})(I - \Phi_{\xi(l,k)})^T \\
&\leq \bar{d}\Phi_{\xi(l,k)} + I - \Phi_{\xi(l,k)} \\
&= \mathrm{diag}\{1, \ldots, 1, \bar{d}, 1, \ldots, 1\},
\end{aligned}$$

where $\bar{d}$ is the s-th diagonal element if and only if $\mathrm{mod}(l\hbar + k - s, q) = 0$. Hence, one has

$$\Gamma_{\xi(l,k)}\Gamma_{\xi(l,k)}^T \leq \bar{\lambda}I.$$

Similarly, it is clear to see that

$$\Gamma_{\xi(l,k)}\Gamma_{\xi(l,k)}^T \geq \underline{d}\Phi_{\xi(l,k)} + I - \Phi_{\xi(l,k)} \geq \underline{\lambda}I > 0$$

which completes the proof.

Lemma 8.4 *Under Assumptions 8.2–8.3, the following inequality is satisfied for $l \in [2 \; \infty)$ and $k \in [1 \; \hbar]$:*

$$K(l,k)K^T(l,k) \leq \underline{\lambda}^{-1}I. \tag{8.27}$$

Proof *It follows from Lemma 8.3, the expressions of $K(l,k)$ and $\tilde{R}(l,k)$ that*

$$\begin{aligned}
&K^T(l,k)K(l,k) \\
&= \tilde{R}^{-1}(l,k)\Gamma_{\xi(l,k)}M_p(l,k)M_p(l,k)\Gamma_{\xi(l,k)}^T\tilde{R}^{-1}(l,k) \\
&\leq \underline{\lambda}^{-1}\tilde{R}^{-1}(l,k)\Gamma_{\xi(l,k)}M_p(l,k)\Gamma_{\xi(l,k)}^T\Gamma_{\xi(l,k)}M_p(l,k)\Gamma_{\xi(l,k)}^T\tilde{R}^{-1}(l,k) \\
&\leq \underline{\lambda}^{-1}I
\end{aligned}$$

which results in the validity of (8.27) from the linear algebra theory and hence ends the proof.

Theorem 8.2 *Under Assumptions 8.2–8.3, for $l \in [2 \; \infty)$ and $k \in [1 \; \hbar]$, the locally minimal upper bound $M_u(l,k)$ of the filtering error variance satisfies the following inequality*

$$M_u(l,k) \leq m(l,k)I \tag{8.28}$$

with initial constraints $M_u(l,0) \leq m(l,0)I$ for $l \in [1 \; \infty)$ and $M_u(1,k) \leq m(1,k)I$ for $k \in [0 \; \hbar]$, where

$$m(l,k) \triangleq \sum_{i=2}^{l} \alpha_1 \gamma(l - i, k - 1)m(i,0)$$

$$+ \sum_{j=1}^{k} \alpha_2 \gamma(l-2, k-j) m(1,j)$$

$$+ \sum_{i=1}^{l-1} \sum_{j=0}^{k-1} \gamma(l-i-1, k-j-1)\beta \qquad (8.29)$$

in which

$$\alpha_1 \triangleq \bar{a}_1(1+\alpha)\left(1 + \sum_{i=1}^{r} \sigma_{\theta_i} \bar{k}\bar{\lambda}\right),$$

$$\alpha_2 \triangleq \bar{a}_2(1+\alpha^{-1})\left(1 + \sum_{i=1}^{r} \sigma_{\theta_i} \bar{k}\bar{\lambda}\right),$$

$$\beta \triangleq \left(\bar{c}_1(\bar{q}+\epsilon_1\bar{r}^2) + \bar{c}_2(\bar{q}+\epsilon_2\bar{r}^2) + \left(\frac{1}{\epsilon_1} + \frac{\tau^2}{4\epsilon}\right)\frac{\bar{a}_1}{\lambda} + (1+\epsilon)\frac{\tau^2}{4} + \frac{\bar{a}_2}{\epsilon_2\lambda}\right)$$

$$\times \left(1 + \sum_{i=1}^{r} \sigma_{\theta_i} \bar{k}\bar{\lambda}\right) + \sum_{i=1}^{r} \sigma_{\theta_i} \bar{k}\left(\bar{r} + \frac{\tau^2}{4}\right),$$

$$\gamma(0,k) \triangleq \alpha_1 \gamma(0, k-1), \quad \gamma(l,0) \triangleq \alpha_2 \gamma(l-1, 0),$$

$$\gamma(l,k) \triangleq \alpha_1 \gamma(l, k-1) + \alpha_2 \gamma(l-1, k), \quad \gamma(0,0) \triangleq 1.$$

Proof *In view of (8.17), Assumptions 8.2–8.3, and Lemma 8.4, one has*

$$M_p(l,k) \le (1+\alpha)\bar{A}_1(l, k-1) M_u(l, k-1)\bar{A}_1^T(l, k-1)$$

$$+ (1+\alpha^{-1})\bar{A}_2(l-1, k) M_u(l-1, k)\bar{A}_2^T(l-1, k)$$

$$+ \left(\bar{c}_1(\bar{q}+\epsilon_1\bar{r}^2) + \bar{c}_2(\bar{q}+\epsilon_2\bar{r}^2) + \left(\frac{1}{\epsilon_1} + \frac{\tau^2}{4\epsilon}\right)\frac{\bar{a}_1}{\lambda}\right.$$

$$\left. + (1+\epsilon)\frac{\tau^2}{4} + \frac{\bar{a}_2}{\epsilon_2\lambda}\right) I$$

$$\triangleq \bar{M}_p(l,k)$$

which together with (8.20) results in that

$$M_u(l,k) \le \bar{M}_p(l,k) + \sum_{i=1}^{r} \sigma_{\theta_i} K_i \left[\Gamma_{\xi(l,k)}\bar{M}_p(l,k)\Gamma_{\xi(l,k)}^T\right.$$

$$\left. + (\bar{r} + \tau^2/4)I\right] K_i^T. \qquad (8.30)$$

In what follows, the proof of assertion (8.28) is carried out by induction that is divided into two steps. The initial step aims to confirm that (8.28) is true with respect to the first pass ($l = 2$). By setting $(l,k) = (2,1)$, we derive from Assumptions 8.2–8.3, Lemma 8.3, and (8.30) that

$$M_u(2,1) \le \bar{M}_p(2,1) + \sum_{i=1}^{r} \sigma_{\theta_i} K_i \left[\Gamma_{\xi(2,1)}\bar{M}_p(2,1)\Gamma_{\xi(2,1)}^T + (\bar{r} + \tau^2/4)I\right] K_i^T$$

$$\le \bar{a}_1(1+\alpha)\left(1 + \sum_{i=1}^{r} \sigma_{\theta_i} \bar{k}\bar{\lambda}\right) m(2,0)I$$

$$+ \bar{a}_2(1 + \alpha^{-1})\left(1 + \sum_{i=1}^{r} \sigma_{\theta_i} \bar{k}\bar{\lambda}\right) m(1,1)I$$

$$+ \left(\bar{c}_1(\bar{q} + \epsilon_1 \bar{r}^2) + \bar{c}_2(\bar{q} + \epsilon_2 \bar{r}^2) + \left(\frac{1}{\epsilon_1} + \frac{\tau^2}{4\epsilon}\right)\frac{\bar{a}_1}{\bar{\lambda}}\right.$$

$$+ (1 + \epsilon)\frac{\tau^2}{4} + \frac{\bar{a}_2}{\epsilon_2 \underline{\lambda}}\right)\left(1 + \sum_{i=1}^{r} \sigma_{\theta_i} \bar{k}\bar{\lambda}\right)I + \sum_{i=1}^{r} \sigma_{\theta_i} \bar{k}\left(\bar{r} + \frac{\tau^2}{4}\right)I$$

$$= (\alpha_1 m(2,0) + \alpha_2 m(1,1) + \beta)I$$

$$= m(2,1)I.$$

Assuming that $M_u(2,k) \leq m(2,k)I$ holds for a fixed integer $k \in [1 \ \hbar - 1]$, one obtains from (8.30) that

$$M_u(2, k+1) \leq \bar{M}_p(2, k+1) + \sum_{i=1}^{r} \sigma_{\theta_i} K_i \Big[\Gamma_{\xi(2,k+1)} \bar{M}_p(2, k+1)\Gamma^T_{\xi(2,k+1)}$$

$$+ (\bar{r} + \tau^2/4)I\Big]K_i^T$$

$$\leq (\alpha_1 m(2,k) + \alpha_2 m(1, k+1) + \beta)I$$

$$= \alpha_1 \Big[\alpha_1 \gamma(0, k-1)m(2,0) + \sum_{j=1}^{k} \alpha_2 \gamma(0, k-j)m(1,j)$$

$$+ \sum_{j=0}^{k-1} \gamma(0, k-j-1)\beta\Big]I + \alpha_2 m(1, k+1)I + \beta I$$

$$= \Big[\alpha_1 \gamma(0, k)m(2,0) + \sum_{j=1}^{k+1} \alpha_2 \gamma(0, k-j+1)m(1,j)$$

$$+ \sum_{j=0}^{k} \gamma(0, k-j)\beta\Big]I$$

$$= m(2, k+1)I$$

which infers that $M_u(2,k) \leq m(2,k)I$ for all $k \in [1 \ \hbar]$.

Next, the inductive step is to verify the validity of (8.28) for $(l,k) \leq (l_0+1, \hbar)$ with the assumption that the assertion (8.28) holds for $(l,k) \leq (l_0, \hbar)$. It follows from the hypothesis that

$$M_u(l_0 + 1, 1) \leq \bar{M}_p(l_0 + 1, 1) + \sum_{i=1}^{r} \sigma_{\theta_i} K_i \Big[\Gamma_{\xi(l_0+1,1)} \bar{M}_p(l_0 + 1, 1)\Gamma^T_{\xi(l_0+1,1)}$$

$$+ (\bar{r} + \tau^2/4)I\Big]K_i^T$$

$$\leq (\alpha_1 m(l_0 + 1, 0) + \alpha_2 m(l_0, 1) + \beta)I$$

$$= \alpha_1 m(l_0 + 1, 0)I + \alpha_2 \Big[\sum_{i=2}^{l_0} \alpha_1 \gamma(l_0 - i, 0)m(i, 0)$$

$$+ \alpha_2 \gamma(l_0 - 2, 0)m(1,1) + \sum_{i=1}^{l_0-1} \gamma(l_0 - i - 1, 0)\beta\Big] I + \beta I$$

$$= \Big[\sum_{i=2}^{l_0+1} \alpha_1 \gamma(l_0 - i + 1, 0)m(i,0) + \alpha_2 \gamma(l_0 - 1, 0)m(1,1)$$

$$+ \sum_{i=1}^{l_0} \gamma(l_0 - i, 0)\beta\Big] I$$

$$= m(l_0 + 1, 1)I.$$

Furthermore, we assume that (8.28) is valid for $(l, k) \leq (l_0+1, k_0)$ with a given $k_0 \in [1 \ \hbar - 1]$. The same assertion can be derived for $(l, k) \leq (l_0 + 1, k_0 + 1)$ as follows:

$$M_u(l_0 + 1, k_0 + 1)$$

$$\leq \bar{M}_p(l_0 + 1, k_0 + 1) + \sum_{i=1}^{r} \sigma_{\theta_i} K_i \Big[\Gamma_{\xi(l_0+1,k_0+1)}$$

$$\times \bar{M}_p(l_0 + 1, k_0 + 1)\Gamma^T_{\xi(l_0+1,k_0+1)} + (\bar{r} + \tau^2/4)I\Big] K_i^T$$

$$\leq (\alpha_1 m(l_0 + 1, k_0) + \alpha_2 m(l_0, k_0 + 1) + \beta)I$$

$$= \alpha_1 \Big[\sum_{i=2}^{l_0+1} \alpha_1 \gamma(l_0 - i + 1, k_0 - 1)m(i,0) + \sum_{j=1}^{k_0} \alpha_2 \gamma(l_0 - 1, k_0 - j)m(1,j)$$

$$+ \sum_{i=1}^{l_0} \sum_{j=0}^{k_0-1} \gamma(l_0 - i, k_0 - j - 1)\beta \Big] I + \alpha_2 \Big[\sum_{i=2}^{l_0} \alpha_1 \gamma(l_0 - i, k_0)m(i,0)$$

$$+ \sum_{j=1}^{k_0+1} \alpha_2 \gamma(l_0 - 2, k_0 - j + 1)m(1,j)$$

$$+ \sum_{i=1}^{l_0-1} \sum_{j=0}^{k_0} \gamma(l_0 - i - 1, k_0 - j)\beta \Big] I + \beta I$$

$$= \sum_{i=2}^{l_0} \alpha_1 \left[\alpha_1 \gamma(l_0 - i + 1, k_0 - 1) + \alpha_2 \gamma(l_0 - i, k_0) \right] m(i,0)I$$

$$+ \alpha_1^2 \gamma(0, k_0 - 1)m(l_0 + 1, 0)I$$

$$+ \sum_{j=1}^{k_0} \alpha_2 [\alpha_1 \gamma(l_0 - 1, k_0 - j) + \alpha_2 \gamma(l_0 - 2, k_0 - j + 1)]$$

$$\times m(1,j)I + \alpha_2^2 \gamma(l_0 - 2, 0)m(1, k_0 + 1)I$$

$$+ \sum_{i=1}^{l_0-1} \sum_{j=0}^{k_0-1} [\alpha_1 \gamma(l_0 - i, k_0 - j - 1) + \alpha_2 \gamma(l_0 - i - 1, k_0 - j)] \beta I + \beta I$$

$$+ \Big[\sum_{j=0}^{k_0-1} \alpha_1\gamma(0, k_0 - j - 1) + \sum_{i=1}^{l_0-1} \alpha_2\gamma(l_0 - i - 1, 0) \Big]\beta I$$

$$= \sum_{i=2}^{l_0+1} \alpha_1\gamma(l_0 - i + 1, k_0)m(i, 0)I$$

$$+ \sum_{j=1}^{k_0+1} \alpha_2\gamma(l_0 - 1, k_0 - j + 1)m(1, j)I + \sum_{i=1}^{l_0}\sum_{j=0}^{k_0} \gamma(l_0 - i, k_0 - j)\beta I$$

$$= m(l_0 + 1, k_0 + 1)I.$$

Based on the inductive method, the proof of this theorem is completed.

Remark 8.7 *Owing to the energy constraint in practical situations, it is natural to put certain limitations on the system parameters. Then, under Assumptions 8.2–8.3, Theorem 8.2 presents the boundedness analysis of the locally minimal upper bound, where an upper bound function is provided for $M_u(l, k)$ and thus for $P_u(l, k)$ as well. Such an upper bound function depends on the information of all the system parameters, noise variances, initial conditions, and RR protocol scheduling.*

Remark 8.8 *In this chapter, one of the novelties lies in that the resilient filter is designed for the considered LRP to ensure the locally minimal upper bound, where the upper bound and gain parameter can be recursively calculated by solving the Riccati-like equations (8.17)–(8.18). Furthermore, the boundedness analysis is discussed with regard to the filtering error variance. It is worth mentioning that we have made great effort to handle the two-directional propagation, shift-varying parameters, RR protocol, and uniform quantization when designing the filter scheme and analyzing the boundedness issue, which accounts for the primary complexity/challenge in this chapter.*

8.3 Numerical Example

In this section, a simulation example is presented to illustrate the usefulness of the designed resilient filter strategy. Consider a shift-varying LRP (8.1)–(8.2) with the following parameters:

$$\vec{A}_1(l, k) = \begin{bmatrix} 0.8 & 0.1\sin(l)\cos(k) \\ 0.1 + 0.05\sin(k) & 0.5 \end{bmatrix},$$

$$\vec{A}_2(l, k) = \begin{bmatrix} 0.55 & 0.1\sin(2k) \\ 0.4 & 0.35 - 0.1e^{-2l} \end{bmatrix},$$

$$\vec{B}_1(l, k) = \begin{bmatrix} 0.5 & 0.2 \\ 0.1\sin(l) & 0.4 \end{bmatrix},$$

$$\vec{B}_2(l, k) = \begin{bmatrix} -0.7 & 0.15\cos(3k) \\ 0.2e^{-2l} & 0.5 \end{bmatrix},$$

$$\vec{C}_1(l, k) = \begin{bmatrix} 0.1 + 0.05\cos(l) & 0.15 \end{bmatrix}^T,$$

$$\vec{C}_2(l, k) = \begin{bmatrix} 0.18 & 0.2 - 0.1e^{-2k} \end{bmatrix}^T,$$

$$\vec{D}_1(l, k) = \begin{bmatrix} 0.6 & 0.3 - 0.15e^{-2k} \\ 0.8 & 0.45 \\ 0.8 & 0.45 \end{bmatrix},$$

$$\vec{D}_2(l, k) = \begin{bmatrix} 0.5 & 0.2 \\ 0.5 + 0.1\sin(l) & 0.6 \\ 0.5 & 0.2 \end{bmatrix}.$$

The pass length is set as $\hbar = 30$. The initial values of (8.1)–(8.2) are given by $d(l+1) = f(k) = [0\ 0]^T$ for $l \in [0\ \infty)$ and $k \in [0\ \hbar]$. The process noises in the first pass and at the start of each pass are supposed to be $w(l+1, 0) = w(1, k) = 0$. Then, it is easy to obtain the boundary condition concerning the transformed FM-II model (8.4) being $x(l, 0) = x(1, k) = [0\ 0\ 0\ 0]^T$ for $l \in [1\ \infty)$ and $k \in [0\ \hbar]$.

For $l \in [2\ \infty)$ and $k \in [1\ \hbar]$, the process and measurement noises follow two uncorrelated Gaussian distributions with zero mean and respective variances $Q(l, k) = 0.16$ and $R(l, k) = 0.25I$. The quantization level is set as $\tau = 0.1$. The gain perturbation processes the form $\Delta(l, k) = \theta_1(l, k)K_1$, where $\theta_1(l, k)$ is a zero-mean Gaussian sequence with variance $\sigma_{\theta_1} = 0.04$ and

$$K_1^T = \begin{bmatrix} 0.15 & -0.02 & 0.1 & 0 & 0.13 & 0.02 & -0.1 \\ 0 & 0.1 & 0.05 & 0.2 & 0.1 & 0 & 0.1 \\ 0.1 & 0 & -0.1 & 0.1 & 0 & 0.1 & 0.05 \end{bmatrix}.$$

In addition, the scaling scalars in this example are chose as $\alpha = 1$, $\epsilon = \epsilon_1 = \epsilon_2 = 1$, and $\mu = 0.25$.

Denote $\bar{x}^{(i)}(l, k)$, $\hat{x}_u^{(i)}(l, k)$, and $e_u^{(i)}(l, k)$ as the i-th elements of $\bar{x}(l, k)$, $\hat{x}_u(l, k)$, and $e_u(l, k)$, respectively. According to the developed theoretical results, the system states can be estimated and the locally minimal upper bound on the filtering error variance is obtained at each step. For space consideration, taking the first element of $\bar{x}(l, k)$ into account, Figs. 8.3–8.5 show the trajectories of $\bar{x}^{(1)}(l, k)$, its estimate $\hat{x}_u^{(1)}(l, k)$, and the corresponding estimation error $e_u^{(1)}(l, k)$, respectively. It can be seen from Figs. 8.3 to 8.5 that the proposed filter performs quite well. Furthermore, Fig. 8.6 plots the trace of the actual one based on 100 independent experimental trials, and Fig. 8.7 presents the trace of the derived upper bound on the filtering error variance. Figures 8.6–8.7 confirm the achievement of the design goal.

To further evaluate the influence of the gain perturbation on the filtering performance, we reset $\sigma_{\theta_1} = 0.4$ while remaining all the other parameters. In this scenario, the trace of the locally minimal upper bound $M_u(l, k)$ is presented in Fig. 8.8. Comparing with Fig. 8.7 and Fig. 8.8, we can see that the upper bound becomes larger as the value of σ_{θ_1} increases.

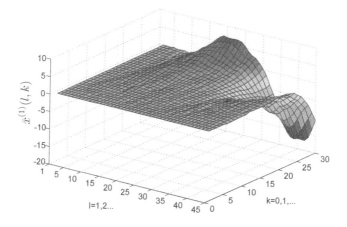

FIGURE 8.3: State evolution of $\bar{x}^{(1)}(l,k)$.

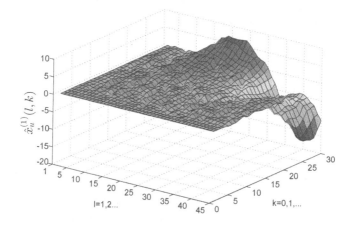

FIGURE 8.4: Estimate evolution of $\hat{x}_u^{(1)}(l,k)$.

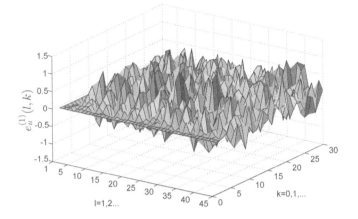

FIGURE 8.5: Estimation error $e_u^{(1)}(i,j)$.

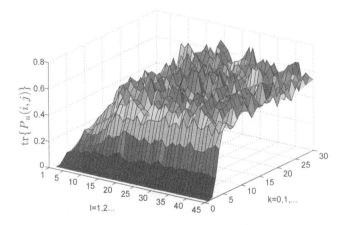

FIGURE 8.6: Trajectory of the actual error variance $P_u(l,k)$.

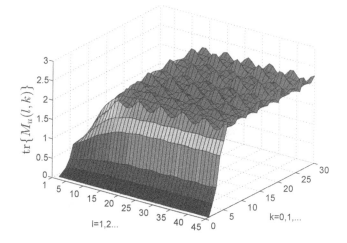

FIGURE 8.7: Trace trajectory of the upper bound $M_u(l,k)$ with $\sigma_{\theta_1} = 0.04$.

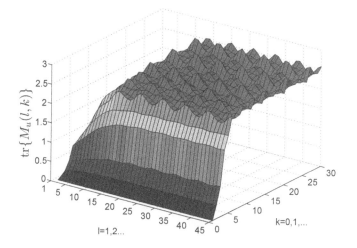

FIGURE 8.8: Trace trajectory of the upper bound $M_u(l,k)$ with $\sigma_{\theta_1} = 0.4$.

8.4 Summary

The resilient filtering problem has been investigated for the linear shift-varying repetitive processes with uniform quantization and RR protocol. The proposed scheduling is implemented in a token-dependent way, where only one sensor is permitted to obtain the network access at certain instant so as to reduce the bandwidth usage. Owing to the quantization error, the accurate estimation error variance is unavailable, not to mention the design of the resilient filter. Therefore, an upper bound on the filtering error variance has been guaranteed at each iteration as an alternative method. Subsequently, the resilient filter has been designed to locally optimize the obtained upper bound. Furthermore, the boundedness of the filtering error variance has been analyzed to reflect the filer performance. A simulation example has also been given to show effectiveness of the designed filter scheme.

9

Event-Triggered Recursive Filtering for Shift-Varying Linear Repetitive Processes

As mentioned before, the LRPs are described by a sequence of sweeps, termed as batches/passes, through specific difference or differential dynamic equations over a finite duration. The corresponding process output of each pass, named as the pass profile, is generated and serves as a forcing function on the forthcoming pass profile. An essential characteristic of the LRPs is that the pass profile displays from pass to pass while the state evolves along the pass subject to a finite duration, which constitutes the intrinsic bidirectional feature.

Concerning the LRPs, the filtering matters have been gaining a recurring research interest in the past few years [18, 90, 228]. Although fruitful outcomes have been built for the filtering problem of LRPs, these results are far from complete and lack systematic investigation on recursive filter design schemes for LRPs, and this is mainly due to the mathematical difficulty in (i) coping with the complex dynamics of repetitive process and (ii) solving out the tightest upper bound on the filtering error variance. On the other hand, most existing results have been concentrated on shift-invariant systems and the relevant application scope tends to be limited since almost all practical systems are shift-dependent. Notice that the products of repetitive processes in industry are certainly implemented in finite quantities over a finite time. On account of the finite duration, concentration of the LRPs is expected to monitor the transient performances rather than regulate the steady-state matters. It hence makes more sense to consider the recursive filtering issue for shift-dependent/varying LRPs with focus on the transient behaviors of error dynamics, which brings about the primary motivation of this study.

It should be stressed that the information exchange executed all the time would result in the broadcast of redundant or unnecessary signals, thereby leading to energy waste. Put another way, the conventional time-triggered tactic with uninterrupted communication depletes the restricted energy cost, and this will further destroy the communication quality. In this scenario, the energy-efficient filtering scheme has become a focus of research in both academia and industry, and the event-based scheduling has emerged naturally for filter design purposes, see [121, 178, 240, 254]. The basic ideology of the event-based mechanism is to first distinguish the valuable signals from massive information and then transmit them only at triggering instants, thereby diminishing the execution frequency meanwhile sustaining a desirable estimation performance. Despite the growing research attention drawn onto the filtering/estimation problems for 1-D systems with event-triggered strategies, the corresponding results have not been considered for LRPs yet, much less the appearance of varying parameters and error variance constraints. This promotes the second motivation for considering the current study.

Following the above discusses, we conclude that there are both theoretical motivation and practical need to launch an investigation on the event–triggered filtering problem for shift-varying LRPs. Nevertheless, some technical challenges we are facing are summarized as follows: (1) how to define an event-triggered shift sequence for repetitive process systems in view of the bidirectional signal propagations? (2) how to obtain certain criterion for deriving a kind of upper bounds on the filtering error covariance based on novel approaches? (3) how to determine a filter in a recursive manner with an event-triggered communication protocol such that the derived upper bound can be optimized at each step? and (4) how to examine the influence of the event threshold on the filtering performance? The main purpose of this chapter is hence to solve the proposed filter design problem by handling the identified difficulties mentioned above.

In this chapter, we endeavor to solve the recursive filter design problem for a class of shift-varying LRPs with an event-triggered scheme. The considered LRPs are first converted into a general FM-II model and then a recursive filter is adopted where the innovation information is updated based on a specified event-triggered rule. The primary novelties are mainly threefold:

(1) An event-triggered strategy is proposed for the shift-varying LRPs to reduce the communication burden. A novel definition of triggering-shift sequence and an event-triggered rule are put forward for the considered system, from which the occurrence of certain events can be determined.

(2) The desired recursive filter is established to find the optimization of certain upper bound in the positive definite sense that always provides a restriction for the estimation error variance. A criterion is exhibited for deriving the expected bound as well as the filter gain based on two series of recursive Riccati-like difference equations.

(3) Elegant discussion is made about the impact of the proposed event-triggered schedule on the filtering performance. Particularly, the monotonicity of the filtering performance regarding the event-triggered threshold is delicately analyzed with rigorous mathematical analysis.

An outline of this chapter is given as follows. Section 9.1 introduces the considered repetitive processes which are then converted to a general FM-II model. Subsequently, the event-triggered filtering problem is formulated for the transformed system. Section 9.2 provides the major results with development of the local minimum upper bound and discussion of the filtering performance. Section 9.3 shows an illustrative simulation and Section 9.4 draws the conclusions.

9.1 Problem Formulation

9.1.1 Linear Repetitive Process

Consider the following linear shift-varying repetitive process:

$$\vec{y}(l+1, k+1) = \vec{A}_1(l+1, k)\vec{y}(l+1, k) + \vec{C}_1(l, k)\vec{v}(l, k)$$
$$+ \vec{B}_1(l+1, k)w(l+1, k), \qquad (9.1a)$$
$$\vec{v}(l+1, k) = \vec{A}_2(l+1, k)\vec{y}(l+1, k) + \vec{C}_2(l, k)\vec{v}(l, k)$$
$$+ \vec{B}_2(l+1, k)w(l+1, k) \qquad (9.1b)$$

where, for $l \in [0 \ \infty)$ and $k \in [0 \ \alpha]$, α is the pass length, $\vec{y}(l+1, k) \in \mathbb{R}^n$ is the state vector, $\vec{v}(l, k) \in \mathbb{R}^m$ is the pass profile vector, and $w(l+1, k) \in \mathbb{R}^p$ is the white process noise with mean zero and covariance $Q(l+1, k)$. Matrices $\vec{A}_\flat(l+1, k)$, $\vec{B}_\flat(l+1, k)$, and $\vec{C}_\flat(l, k)$ ($\flat = 1, 2$) are known, shift-varying ones with appropriate dimensions.

Note that the present state $\vec{y}(l+1, k)$ and the pass profile $\vec{v}(l, k)$ are often inaccessible directly. The available measurement defined on $l \in [0 \ \infty)$ and $k \in [0 \ \alpha]$ is expressed as follows:

$$z(l+1, k) = \vec{E}_1(l+1, k)\vec{y}(l+1, k) + \vec{E}_2(l, k)\vec{v}(l, k) + \eta(l+1, k) \qquad (9.2)$$

where $z(l+1, k) \in \mathbb{R}^q$ is the observed output, $\eta(l+1, k) \in \mathbb{R}^q$ signifying the measurement noise is a zero-mean white sequence with covariance $R(l+1, k) > 0$, and $\vec{E}_\flat(l, k)$ ($\flat = 1, 2$) are known shift-varying matrices. The initial states associated with (9.1) are given by

$$\mathbb{E}\{\vec{y}(l+1, 0)\} = f(l+1), \quad \mathbb{E}\{\vec{v}(0, k)\} = g(k) \qquad (9.3)$$

where $f(l+1) \in \mathbb{R}^n$ and $g(k) \in \mathbb{R}^m$ are known vectors.

Throughout this chapter, the following assumption is given.

Assumption 9.1 *For any $l_\hbar$ and $k_\hbar$ ($\hbar \in [1 \ 3]$), the initial states $\vec{y}(l_1, 0)$ and $\vec{v}(0, k_1)$, the measurement noise $\eta(l_2, k_2)$, and the process noise $w(l_3, k_3)$ are assumed to be uncorrelated with each other.*

To facilitate later analysis, the underlying system is to be transformed into a general FM-II state-space model. By defining

$$y(l+1, k) = \begin{bmatrix} \vec{y}^T(l+1, k) & \vec{v}^T(l, k) \end{bmatrix}^T,$$
$$A_1(l+1, k) = \begin{bmatrix} \vec{A}_1(l+1, k) & \vec{C}_1(l, k) \\ 0 & 0 \end{bmatrix},$$
$$A_2(l, k+1) = \begin{bmatrix} 0 & 0 \\ \vec{A}_2(l, k+1) & \vec{C}_2(l-1, k+1) \end{bmatrix},$$

$$B_1(l+1,k) = \begin{bmatrix} \vec{B}_1(l+1,k) \\ 0 \end{bmatrix},$$

$$B_2(l,k+1) = \begin{bmatrix} 0 \\ \vec{B}_2(l,k+1) \end{bmatrix},$$

$$E(l,k) = \begin{bmatrix} \vec{E}_1(l,k) & \vec{E}_2(l-1,k) \end{bmatrix}$$

we obtain from (9.1) to (9.2) that

$$\begin{cases} y(l+1,k+1) = A_1(l+1,k)y(l+1,k) + A_2(l,k+1)y(l,k+1) \\ \qquad\qquad + B_1(l+1,k)w(l+1,k) + B_2(l,k+1)w(l,k+1), \quad (9.4) \\ z(l,k) = E(l,k)y(l,k) + \eta(l,k) \end{cases}$$

for $l \in [1 \ \infty)$ and $k \in [0 \ \ \alpha-1]$. The initial boundary condition for (9.4) is presented as

$$y(l,0) = \begin{bmatrix} \vec{y}(l,0) \\ \vec{v}(l-1,0) \end{bmatrix}, \quad l \in [1 \ \infty) \qquad (9.5a)$$

$$y(1,k) = \begin{bmatrix} \vec{y}(1,k) \\ \vec{v}(0,k) \end{bmatrix}, \quad k \in [0 \ \alpha] \qquad (9.5b)$$

in which $\mathbb{E}\{\vec{y}(1,k)\}$ and $\mathbb{E}\{\vec{v}(l-1,0)\}$ can be calculated from (9.1) and (9.3).

Remark 9.1 *Note that the typical repetitive process is featured by a succession of sweeps termed batches/passes, whose dynamic behaviors are defined over certain finite duration. The process output associated with each pass, termed the pass profile, can be acquired and regarded as a forcing function reacting on the upcoming pass profile. Such a process has been broadly regarded as a particular kind of 2-D systems, where the state evolution follows each batch/pass with a certain fixed length, and propagation of the pass profile follows the pass-to-pass orientation. In this case, the FM-II paradigm is naturally applied to investigate the recursive filtering theme of the considered LRP through some novel and compelling methods.*

9.1.2   Event-Triggered Mechanism

The recursive filtering problem will be delicately handled for system (9.1)–(9.2) (or the transformed system (9.4)).

A classical 2-D Kalman-type filter for (9.4) has the following structure:

$$\hat{y}_p(l,k) = A_1(l,k-1)\hat{y}_u(l,k-1) + A_2(l-1,k)\hat{y}_u(l-1,k), \qquad (9.6a)$$

$$\hat{y}_u(l,k) = \hat{y}_p(l,k) + K(l,k)\left(z(l,k) - E(l,k)\hat{y}_p(l,k)\right) \qquad (9.6b)$$

in which, for $l \in [2 \ \infty)$ and $k \in [1 \ \alpha]$, $\hat{y}_p(l,k)$ and $\hat{y}_u(l,k)$ denote the prediction and updated estimate of state $y(l,k)$, respectively, and $K(l,k)$ is the gain parameter in filter (9.6).

A traditional 2-D filter design problem can be investigated from (9.6) by broadcasting the received measurement $z(l, k)$ at each shifting step (l, k). Unfortunately, the periodic or consecutive communication would inevitably lead to the transmission of unimportant/unnecessary data, thereby increasing the network burden and the energy consumption. Rather than the time-scheduled scheme, the event-triggered procedure is to be introduced here when designing the filter structure so as to enhance the utilization efficiency of the network communication. In the following, an event-triggered rule will be proposed to discern when the observed data shall be transmitted to the filter.

To start with, let us define an ordered sequence with two indexes. The notation $(l_1, k_1) < (l_2, k_2)$ indicates $(l_1, k_1) \in \{(l, k) | l = l_2, k < k_2\} \cup \{(l, k) | l < l_2, k \in [1\,\alpha]\}$, and the equality $(l_1, k_1) = (l_2, k_2)$ holds if and only if $l_1 = l_2$ and $k_1 = k_2$. Based on such a definition, an event triggering sequence is denoted as

$$(2, 1) = (l_1, k_1) < \cdots < (l_t, k_t) < \cdots,$$

where (l_t, k_t) $(t \in [1 \, \infty))$ represents the t-th event-triggered instant.

Denote $e(l_t, k_t) \triangleq z(l_t, k_t) - z(l, k)$ with $z(l_t, k_t)$ being the measurement received at (l_t, k_t). It is noteworthy that the latest event execution does not perform at the t-th triggering instant until the following inequality is satisfied

$$f(e(l_t, k_t), \delta) > 0, \tag{9.7}$$

where $f(e(l_t, k_t), \delta) \triangleq e^T(l_t, k_t) e(l_t, k_t) - \delta$ is the event generator function and δ is a given positive scalar. In this case, the $(t + 1)$-th triggering-shift step is determined by

$$(l_{t+1}, k_{t+1}) = \min\{(l, k) | (l, k) > (l_t, k_t), f(e(l_t, k_t), \delta) > 0\}, \tag{9.8}$$

where $\min\{(l, k) | \text{Condition 1}, \text{Condition 2}, \ldots, \text{Condition M}\}$ represents the first shift step in the ordered sequence satisfying all the M conditions.

Remark 9.2 *In the light of (9.7), it is evident to see that the triggering frequency is dependent on both the measurement error $e(l_t, k_t)$ and a certain threshold δ. Under the condition that the event generator function overweighs zero, the latest measurement is regarded as valuable information that should be applied to update the estimate such that an acceptable filter performance could be guaranteed for the target plant. Note that a tighter threshold would result in a higher triggering frequency. With a smaller value of δ, a better filtering performance could be expected whereas unnecessary data transmissions may increase. Particularly, the event-triggered strategy comes down to the conventional time-driven one if the threshold is set as $\delta = 0$.*

For $(l_t, k_t) \leq (l, k) < (l_{t+1}, k_{t+1})$, a 2-D event-triggered recursive filter is adopted in the following form:

$$\hat{y}_p(l, k) = A_1(l, k - 1)\hat{y}_u(l, k - 1) + A_2(l - 1, k)\hat{y}_u(l - 1, k), \tag{9.9a}$$

$$\hat{y}_u(l,k) = \hat{y}_p(l,k) + \mathcal{K}(l,k)\left(z(l_t,k_t) - E(l,k)\hat{y}_p(l,k)\right) \tag{9.9b}$$

with $\mathcal{K}(l,k)$ being the expected gain parameter for $l \in [2\ \infty)$ and $k \in [1\ \alpha]$. The initial conditions for (9.9) are set as $\hat{y}_u(l,0) = \mathbb{E}\{y(l,0)\}$ and $\hat{y}_u(1,k) = \mathbb{E}\{y(1,k)\}$ ($l \in [1\ \infty)$, $k \in [0\ \alpha]$).

Remark 9.3 *It is noticeable that the proposed event-triggered mechanism relies on two indices due to the bidirectional evolution of the consider LRP. Compared with other traditional communication protocols in the literature, the adopted scheduling has identified novelties lying in that a novel two-index-based sequence is first introduced and an exquisite updating rule is subsequently presented for the event execution. The filter strategy under consideration is implemented by the following procedure. To begin with, the first event step is set as $(l_1, k_1) = (2, 1)$ to initialize the triggering-shift sequence. Then, the measurement is observed from the sensor, and the proposed triggering rule is utilized to judge whether an event occurs or not at each shifting step (l, k) ($l \in [2\ \infty)$, $k \in [1\ \alpha]$). The sensor sends out the latest measurement to the filter (9.9) to update the estimate if and only if condition (9.7) is fulfilled. The 2-D event-triggered filtering procedure operates for (9.4) by successively repeating such a bidirectional process.*

Let the prediction error be $\tilde{y}_p(l,k) \triangleq y(l,k) - \hat{y}_p(l,k)$ and the estimation error be $\tilde{y}_u(l,k) \triangleq y(l,k) - \hat{y}_u(l,k)$. Then, it follows from (9.4) and (9.9) that

$$\begin{aligned}
\tilde{y}_p(l,k) &= A_1(l,k-1)\tilde{y}_u(l,k-1) + A_2(l-1,k)\tilde{y}_u(l-1,k) \\
&\quad + B_1(l,k-1)w(l,k-1) + B_2(l-1,k)w(l-1,k), \tag{9.10a} \\
\tilde{y}_u(l,k) &= (I - \mathcal{K}(l,k)E(l,k))\tilde{y}_p(l,k) \\
&\quad - \mathcal{K}(l,k)e(l_t,k_t) - \mathcal{K}(l,k)\eta(l,k). \tag{9.10b}
\end{aligned}$$

Our objective is formulated as ascertaining a filter (9.9) with appropriate gain matrix $\mathcal{K}(l,k)$ such that (1) the estimation error variance $\mathbb{E}\{\tilde{y}_u(l,k)\tilde{y}_u^T (l,k)\}$ has an upper bound and (2) the minimal bound is attained in the positive definite sense at each step.

9.2   Main Results

In what follows, the recursive filtering problem is to be solved for the transformed system (9.4). An upper bound as the estimation performance will be firstly constructed based on the prediction and estimation error variances, and then the obtained bound will be optimized at every iteration by appropriately electing the filter parameter. Subsequently, monotonicity of the minimal upper bound is to be discussed with regard to the event threshold.

For the sake of convenience, we denote

$$\mathcal{P}_p(l, k) \triangleq \mathbb{E}\left\{\tilde{y}_p(l, k)\tilde{y}_p^T(l, k)\right\},$$
$$\mathcal{P}_u(l, k) \triangleq \mathbb{E}\left\{\tilde{y}_u(l, k)\tilde{y}_u^T(l, k)\right\}.$$

The dynamical evolutions of $\mathcal{P}_p(l, k)$ and $\mathcal{P}_u(l, k)$ are provided in the following lemma.

Lemma 9.1 *The recursions of matrices* $\mathcal{P}_p(l, k)$ *and* $\mathcal{P}_u(l, k)$ *satisfy the following Riccati-like equations:*

$$\begin{aligned}
\mathcal{P}_p(l, k) &= A_1(l, k-1)\mathcal{P}_u(l, k-1)A_1^T(l, k-1) \\
&\quad + A_2(l-1, k)\mathcal{P}_u(l-1, k)A_2^T(l-1, k) \\
&\quad + A_1(l, k-1)\mathbb{E}\left\{\tilde{y}_u(l, k-1)\tilde{y}_u^T(l-1, k)\right\}A_2^T(l-1, k) \\
&\quad + A_2(l-1, k)\mathbb{E}\left\{\tilde{y}_u(l-1, k)\tilde{y}_u^T(l, k-1)\right\}A_1^T(l, k-1) \\
&\quad + B_1(l, k-1)Q(l, k-1)B_1^T(l, k-1) \\
&\quad + B_2(l-1, k)Q(l-1, k)B_2^T(l-1, k), \quad\quad\quad (9.11) \\
\mathcal{P}_u(l, k) &= (I - \mathcal{K}(l, k)E(l, k))\mathcal{P}_p(l, k)(I - \mathcal{K}(l, k)E(l, k))^T \\
&\quad + \mathcal{K}(l, k)\left(\mathbb{E}\left\{e(l_t, k_t)e^T(l_t, k_t)\right\} + R(l, k)\right)\mathcal{K}^T(l, k) \\
&\quad + (I - \mathcal{K}(l, k)E(l, k))\mathbb{E}\left\{\tilde{y}_p(l, k)e^T(l_t, k_t)\right\}\mathcal{K}^T(l, k) \\
&\quad + \mathcal{K}(l, k)\mathbb{E}\left\{e(l_t, k_t)\tilde{y}_p^T(l, k)\right\}(I - \mathcal{K}(l, k)E(l, k))^T \\
&\quad + \mathcal{K}(l, k)\left(\mathbb{E}\left\{e(l_t, k_t)\eta^T(l, k)\right\}\right. \\
&\quad \left. + \mathbb{E}\left\{\eta(l, k)e^T(l_t, k_t)\right\}\right)\mathcal{K}^T(l, k). \quad\quad\quad (9.12)
\end{aligned}$$

Proof *Note that* $\tilde{y}_u(l, k)$ *is uncorrelated with* $w(i, j)$ *for* $(i, j) \in \{(i_0, j_0)|i_0 > l \text{ or } j_0 > k\} \cup (l, k)$, *and* $\tilde{y}_p(l, k)$ *is uncorrelated with* $\eta(l, k)$. *The assertion of Lemma 9.1 follows directly from (9.10) and the definitions of* $\mathcal{P}_p(l, k)$ *and* $\mathcal{P}_u(l, k)$, *where the detailed proof is omitted here for brevity.*

Remark 9.4 *On account of the error* $e(l_t, k_t)$ *induced by the event-triggered schedule, the estimation error variance cannot be analytically calculated, let alone the design of the filter parameter that minimizes the error variance. To handle such a challenging issue, we intend to construct an upper bound substituting for the unavailable error variance and then determine the tightest one with appropriate gain parameters.*

Before establishing the main results, we will further discuss the cross term $\mathbb{E}\left\{e(l_t, k_t)\eta^T(l, k)\right\}$ involved in (9.12). An indicator variable $\gamma(l, k)$ taking values on 1 or 0 is introduced to better depict the event triggered scheme. To be more specific, $\gamma(l, k) = 0$ means that the event-triggered mechanism (9.7) is fulfilled at the current shift step (l, k), while $\gamma(l, k) = 1$ represents non-occurrence of the triggering event.

Recalling the definition of $e(l_t, k_t)$, one has $e(l_t, k_t) = 0$ in the case that $(l, k) = (l_t, k_t)$ is an event triggering instant. It is easy to derive

$$
\begin{aligned}
&\mathbb{E}\left\{e(l_t, k_t)\eta^T(l, k)\right\} \\
&= \mathbb{E}\left\{(z(l_t, k_t) - z(l, k))\eta^T(l, k)\right\} \\
&= \mathbb{E}\left\{[z(l_t, k_t) - (E(l, k)y(l, k) + \eta(l, k))]\eta^T(l, k)\right\} \\
&= -\gamma(l, k)R(l, k) \\
&\leq 0.
\end{aligned}
\tag{9.13}
$$

In addition, the event-triggered mechanism results in

$$
e^T(l_t, k_t)e(l_t, k_t) \leq \delta
\tag{9.14}
$$

for any $(l_t, k_t) \in \{(l_s, k_s)\}_{s=1}^\infty$, which indicates

$$
\mathbb{E}\{e(l_t, k_t)e^T(l_t, k_t)\} \leq \delta I.
\tag{9.15}
$$

According to the fundamental matrix inequality, we apparently obtain that

$$
\begin{aligned}
\mathcal{P}_p(l, k) \leq{}& (1 + \mu)A_1(l, k-1)\mathcal{P}_u(l, k-1)A_1^T(l, k-1) \\
&+ (1 + \mu^{-1})A_2(l-1, k)\mathcal{P}_u(l-1, k)A_2^T(l-1, k) \\
&+ B_1(l, k-1)Q(l, k-1)B_1^T(l, k-1) \\
&+ B_2(l-1, k)Q(l-1, k)B_2^T(l-1, k)
\end{aligned}
\tag{9.16}
$$

for any given scalar $\mu > 0$. Furthermore, substituting (9.13) and (9.15) into (9.12) yields

$$
\begin{aligned}
\mathcal{P}_u(l, k) \leq{}& (I - \mathcal{K}(l, k)E(l, k))\mathcal{P}_p(l, k)(I - \mathcal{K}(l, k)E(l, k))^T \\
&+ \mathcal{K}(l, k)(\delta I + R(l, k))\mathcal{K}^T(l, k) \\
&+ (I - \mathcal{K}(l, k)E(l, k))\mathbb{E}\left\{\tilde{y}_p(l, k)e^T(l_t, k_t)\right\}\mathcal{K}^T(l, k) \\
&+ \mathcal{K}(l, k)\mathbb{E}\left\{e(l_t, k_t)\tilde{y}_p^T(l, k)\right\}(I - \mathcal{K}(l, k)E(l, k))^T \\
\leq{}& (1 + \epsilon)(I - \mathcal{K}(l, k)E(l, k))\mathcal{P}_p(l, k)(I - \mathcal{K}(l, k)E(l, k))^T \\
&+ \mathcal{K}(l, k)\left((1 + \epsilon^{-1})\delta I + R(l, k)\right)\mathcal{K}^T(l, k)
\end{aligned}
\tag{9.17}
$$

for any given scalar $\epsilon > 0$.

The following theorem successfully constructs the upper bounds of matrices $\mathcal{P}_p(l, k)$ and $\mathcal{P}_u(l, k)$.

Theorem 9.1 *Let μ, ϵ be given positive scalars. For $l \in [2\ \infty)$ and $k \in [1\ \alpha]$, if there exist two sequences of positive definite matrices $\{M_p(l, k)\}$ and $\{M_u(l, k)\}$ satisfying the following Riccati-like difference equations*

$$
M_p(l, k) = (1 + \mu)A_1(l, k-1)M_u(l, k-1)A_1^T(l, k-1)
$$

$$+ (1 + \mu^{-1})A_2(l-1,k)M_u(l-1,k)A_2^T(l-1,k)$$
$$+ B_1(l,k-1)Q(l,k-1)B_1^T(l,k-1)$$
$$+ B_2(l-1,k)Q(l-1,k)B_2^T(l-1,k), \tag{9.18}$$
$$M_u(l,k) = (1+\epsilon)(I - \mathcal{K}(l,k)E(l,k))M_p(l,k)(I - \mathcal{K}(l,k)E(l,k))^T$$
$$+ \mathcal{K}(l,k)\left((1+\epsilon^{-1})\delta I + R(l,k)\right)\mathcal{K}^T(l,k) \tag{9.19}$$

with initial conditions

$$M_u(l,0) = \mathcal{P}_u(l,0), \quad M_u(1,k) = \mathcal{P}_u(1,k) \tag{9.20}$$

then the error variance matrices are bounded by

$$\mathcal{P}_p(l,k) \le M_p(l,k), \quad \mathcal{P}_u(l,k) \le M_u(l,k) \tag{9.21}$$

for $l \in [2\ \infty)$ and $k \in [1\ \alpha]$. Moreover, the minimum upper bound $M_u(l,k)$ regarding the error variance satisfies the following recursion

$$M_u(l,k) = (1+\epsilon)M_p(l,k) - (1+\epsilon)^2 M_p(l,k)$$
$$\times E^T(l,k)\bar{R}^{-1}(M_p(l,k))E(l,k)M_p(l,k) \tag{9.22}$$

with filter gain parameter computed by

$$\mathcal{K}(l,k) = (1+\epsilon)M_p(l,k)E^T(l,k)\bar{R}^{-1}(M_p(l,k)) \tag{9.23}$$

where

$$\bar{R}(M_p(l,k)) = (1+\epsilon)E(l,k)M_p(l,k)E^T(l,k) + (1+\epsilon^{-1})\delta I + R(l,k).$$

Proof *The statement in this theorem would be carried out according to the inductive method.*

Firstly, validity of (9.21) is confirmed with $(l,k) = (2,1)$. It follows from (9.16) to (9.20) that

$$\mathcal{P}_p(2,1) - M_p(2,1)$$
$$\le (1+\mu)A_1(2,0)(\mathcal{P}_u(2,0) - M_u(2,0))A_1^T(2,0)$$
$$+ (1+\mu^{-1})A_2(1,1)(\mathcal{P}_u(1,1) - M_u(1,1))A_2^T(1,1)$$
$$\le 0$$

which further indicates

$$\mathcal{P}_u(2,1) - M_u(2,1)$$
$$\le (1+\epsilon)(I - \mathcal{K}(2,1)E(2,1))(\mathcal{P}_p(2,1) - M_p(2,1))(I - \mathcal{K}(2,1)E(2,1))^T$$
$$\le 0.$$

Assume that $\mathcal{P}_u(2,j) \leq M_u(2,j)$ is true for a given integer $j \in [1\ \alpha-1]$. Combining such a hypothesis with (9.18)–(9.20), we obtain

$$\mathcal{P}_p(2,j+1) - M_p(2,j+1)$$
$$\leq (1+\mu)A_1(2,j)(\mathcal{P}_u(2,j) - M_u(2,j))A_1^T(2,j) + (1+\mu^{-1})$$
$$\times A_2(1,j+1)(\mathcal{P}_u(1,j+1) - M_u(1,j+1))A_2^T(1,j+1)$$
$$\leq 0, \tag{9.24}$$
$$\mathcal{P}_u(2,j+1) - M_u(2,j+1)$$
$$\leq (1+\epsilon)(I - \mathcal{K}(2,j+1)E(2,j+1))(\mathcal{P}_p(2,j+1) - M_p(2,j+1))$$
$$\times (I - \mathcal{K}(2,j+1)E(2,j+1))^T$$
$$\leq 0. \tag{9.25}$$

Therefore, the inequality $\mathcal{P}_u(2,k) \leq M_u(2,k)$ is valid for all $k \in [1\ \alpha]$ according to mathematical induction.

Next, assume that $\mathcal{P}_u(l,k) \leq M_u(l,k)$ is true for $(l,k) \leq (i,\alpha)$, i.e., $(l,k) \in \{(l_0,k_0)|l_0 \leq i, k_0 \in [1\ \alpha]\}$, where $i \in [2\ \infty)$ is a given integer. Then, it is evident to see

$$\mathcal{P}_p(i+1,1) - M_p(i+1,1)$$
$$\leq (1+\mu)A_1(i+1,0)(\mathcal{P}_u(i+1,0) - M_u(i+1,0))A_1^T(i+1,0)$$
$$+ (1+\mu^{-1})A_2(i,1)(\mathcal{P}_u(i,1) - M_u(i,1))A_2^T(i,1)$$
$$\leq 0,$$
$$\mathcal{P}_u(i+1,1) - M_u(i+1,1)$$
$$\leq (1+\epsilon)(I - \mathcal{K}(i+1,1)E(i+1,1))(\mathcal{P}_p(i+1,1) - M_p(i+1,1))$$
$$\times (I - \mathcal{K}(i+1,1)E(i+1,1))^T$$
$$\leq 0.$$

Similar to the derivation of (9.24) and (9.25), the following inequality can be confirmed

$$\mathcal{P}_u(i+1,k) \leq M_u(i+1,k), \quad \forall k \in [1\ \alpha]$$

which signifies that $\mathcal{P}_u(l,k) \leq M_u(l,k)$ is true for $(l,k) \leq (i+1,\alpha)$. Repeating the above procedure with respect to k pass-by-pass, (9.21) is confirmed for all $l \in [2\ \infty)$ and $k \in [1\ \alpha]$, and hence the proof of (9.21) is complete.

It remains now to verify that the recursive filter (9.9) with parameter (9.23) minimizes the obtained matrix $M_u(l,k)$. By employing the completing-the-square technique to (9.19), we easily arrive at

$$M_u(l,k) = (1+\epsilon)\left[M_p(l,k) - \mathcal{K}(l,k)E(l,k)M_p(l,k) - M_p(l,k)\right.$$
$$\times E^T(l,k)\mathcal{K}^T(l,k)\right] + \mathcal{K}(l,k)\left[(1+\epsilon^{-1})\delta I + R(l,k)\right.$$
$$+ (1+\epsilon)E(l,k)M_p(l,k)E^T(l,k)\right]\mathcal{K}^T(l,k)$$

$$= (1+\epsilon)M_p(l,k) + \left[\mathcal{K}(l,k) - (1+\epsilon)M_p(l,k)E^T(l,k)\bar{R}^{-1}(M_p(l,k))\right]$$
$$\times \bar{R}(M_p(l,k))\left[\mathcal{K}(l,k) - (1+\epsilon)M_p(l,k)E^T(l,k)\bar{R}^{-1}(M_p(l,k))\right]^T$$
$$- (1+\epsilon)^2 M_p(l,k)E^T(l,k)\bar{R}^{-1}(M_p(l,k))E(l,k)M_p(l,k). \tag{9.26}$$

Therefore, the desired filter parameter is selected as (9.23) to achieve the optimal upper bound described by (9.22), which ends the proof.

Notice that the event threshold δ is involved in, and thus contributes to, the evolution of the optimal upper bound $M_u(l,k)$ of the estimation error variance. It would be interesting to consider the effect of the event threshold on $M_u(l,k)$. Intuitively, a smaller threshold infers a higher occurrence rate of event triggering, which would result in a tighter bound due to more available measurements. In what follows, the relationship between the threshold δ and the minimal upper bound $M_u(l,k)$ will be discussed with rigorous analysis.

For presentation convenience, the notations $M_p^\delta(l,k)$ and $M_u^\delta(l,k)$ stand for the minimum upper bounds with a prescribed event threshold δ. In this case, the desired gain parameter (9.23), denoted as $\mathcal{K}_{opt}^\delta(l,k)$, is rewritten by

$$\mathcal{K}_{opt}^\delta(l,k) = (1+\epsilon)M_p^\delta(l,k)E^T(l,k)\bar{R}^{-1}(M_p^\delta(l,k)).$$

Let $X(l,k)$, $Y > 0$ be positive definite matrices and $\delta > 0$ be a given scalar. Define the following two matrix functions:

$$\mathcal{G}(X(l,k-1), X(l-1,k))$$
$$\triangleq (1+\mu)A_1(l,k-1)X(l,k-1)A_1^T(l,k-1)$$
$$+ (1+\mu^{-1})A_2(l-1,k)X(l-1,k)A_2^T(l-1,k)$$
$$+ B_1(l,k-1)Q(l,k-1)B_1^T(l,k-1)$$
$$+ B_2(l-1,k)Q(l-1,k)B_2^T(l-1,k),$$
$$\mathcal{F}((l,k), \delta, \mathcal{K}(l,k), Y)$$
$$\triangleq (1+\epsilon)(I - \mathcal{K}(l,k)E(l,k))Y(I - \mathcal{K}(l,k)E(l,k))^T$$
$$+ \mathcal{K}(l,k)\left((1+\epsilon^{-1})\delta I + R(l,k)\right)\mathcal{K}^T(l,k).$$

Then, it follows from (9.18), (9.19), (9.22), and (9.23) that

$$M_p^\delta(l,k) = \mathcal{G}(M_u^\delta(l,k-1), M_u^\delta(l-1,k)), \tag{9.27}$$
$$M_u^\delta(l,k) = \mathcal{F}((l,k), \delta, \mathcal{K}_{opt}^\delta(l,k), M_p^\delta(l,k)). \tag{9.28}$$

It is obvious to have that $\mathcal{G}(X(l,k-1), X(l-1,k))$ is non-decreasing as $X(l,k-1)$ or $X(l-1,k)$ increases, and $\mathcal{F}((l,k), \delta, \mathcal{K}(l,k), Y)$ is non-decreasing with the increasing of scalar δ or matrix Y. Moreover, according to the results developed in Theorem 9.1, we obtain the following inequality immediately:

$$\mathcal{F}((l,k), \delta, \mathcal{K}_{opt}^\delta(l,k), M_p^\delta(l,k))$$

$$\leq \mathcal{F}((l,k), \delta, \mathcal{K}(l,k), M_p^\delta(l,k)), \quad \forall \mathcal{K}(l,k). \tag{9.29}$$

The following theorem illustrates the influence of the event threshold δ on the optimal upper bound $M_u^\delta(l,k)$.

Theorem 9.2 *For given positive scalars δ_1 and δ_2, if the condition $\delta_1 \leq \delta_2$ holds, then the following relationship*

$$M_u^{\delta_1}(l,k) \leq M_u^{\delta_2}(l,k) \tag{9.30}$$

is true for all $l \in [2 \ \infty)$ and $k \in [1 \ \alpha]$.

Proof *Along a parallel procedure to the reasoning of Theorem 9.1, the assertion of this theorem is proved by resorting to mathematical induction.*

Note the fact that the given initial conditions $M_u(l,0)$ and $M_u(1,k)$ ($l \in [2 \ \infty)$, $k \in [1 \ \alpha]$) are independent of the event threshold. The following expression

$$M_p^\delta(2,1) = \mathcal{G}(M_u(2,0), M_u(1,1))$$

reveals that the event threshold δ has no influence on the matrix $M_p^\delta(2,1)$. In other words, we have

$$M_p^\delta(2,1) = M_p^{\delta_1}(2,1) = M_p^{\delta_2}(2,1). \tag{9.31}$$

In view of (9.28), (9.29), and (9.31), it is easy to derive

$$\begin{aligned}
M_u^{\delta_1}(2,1) &= \mathcal{F}((2,1), \delta_1, \mathcal{K}_{opt}^{\delta_1}(2,1), M_p^{\delta_1}(2,1)) \\
&\leq \mathcal{F}((2,1), \delta_1, \mathcal{K}_{opt}^{\delta_2}(2,1), M_p^{\delta_1}(2,1)) \\
&\leq \mathcal{F}((2,1), \delta_2, \mathcal{K}_{opt}^{\delta_2}(2,1), M_p^{\delta_2}(2,1)) \\
&= M_u^{\delta_2}(2,1).
\end{aligned}$$

Assume that $M_u^{\delta_1}(2,j) \leq M_u^{\delta_2}(2,j)$ is satisfied for a certain integer $j \in [1 \ \alpha - 1]$. By using the monotonicity property of $\mathcal{G}(X(l,k-1), X(l-1,k))$ in $X(l,k-1)$ and $X(l-1,k)$, one has

$$\begin{aligned}
M_p^{\delta_1}(2,j+1) &= \mathcal{G}(M_u^{\delta_1}(2,j), M_u^{\delta_1}(1,j+1)) \\
&= \mathcal{G}(M_u^{\delta_1}(2,j), M_u^{\delta_2}(1,j+1)) \\
&\leq \mathcal{G}(M_u^{\delta_2}(2,j), M_u^{\delta_2}(1,j+1)) \\
&= M_p^{\delta_2}(2,j+1). \tag{9.32}
\end{aligned}$$

Moreover, we obtain from (9.29) and (9.32) that

$$\begin{aligned}
M_u^{\delta_1}(2,j+1) &= \mathcal{F}((2,j+1), \delta_1, \mathcal{K}_{opt}^{\delta_1}(2,j+1), M_p^{\delta_1}(2,j+1)) \\
&\leq \mathcal{F}((2,j+1), \delta_1, \mathcal{K}_{opt}^{\delta_2}(2,j+1), M_p^{\delta_1}(2,j+1))
\end{aligned}$$

$$\leq \mathcal{F}((2,j+1),\delta_1,\mathcal{K}_{opt}^{\delta_2}(2,j+1),M_p^{\delta_2}(2,j+1))$$
$$\leq \mathcal{F}((2,j+1),\delta_2,\mathcal{K}_{opt}^{\delta_2}(2,j+1),M_p^{\delta_2}(2,j+1))$$
$$= M_u^{\delta_2}(2,j+1)$$

where the monotonicity-like property of matrix function $\mathcal{F}(\cdot,\cdot,\cdot,\cdot)$ has been applied in the derivation. Hence, it follows from the induction that

$$M_u^{\delta_1}(2,k) \leq M_u^{\delta_2}(2,k), \quad \forall k \in [1\ \alpha]. \tag{9.33}$$

Proceeding the proof of (9.33) with regard to k pass-by-pass, the validity of (9.30) is verified for all $l \in [2\ \infty)$ and $k \in [1\ \alpha]$, which concludes the proof.

Remark 9.5 *According to Theorem 9.2, it is concluded that the local minimum bound $M_u^{\delta}(l,k)$ is non-decreasing as the threshold value δ increases, which is in accordance with the engineering practice. As previously mentioned, decreasing the threshold represents a larger number of signal transmissions, thereby improving the estimation accuracy. On the other hand, a smaller value of δ leads to more network resource occupancy. Clearly, there is a tradeoff between the energy saving and the filtering performance.*

Remark 9.6 *In Theorem 9.1, for the addressed event-triggered state estimation problem for LRPs, we have constructed an upper bound on the filtering error variance at each step and then minimized such a bound with appropriately designed filter parameters via calculating two series of Riccati-like difference equations. Furthermore, in Theorem 9.2, we have rigorously proved the monotonicity of the filtering performance with regard to the event-triggering threshold. The novelties of our research are highlighted as follows: (1) an event-triggered shifting sequence is defined for repetitive process systems given the bidirectional signal propagations; (2) the existence of certain upper bound is guaranteed on the error covariance and such an upper bound is optimized in the presence of the event-triggered communication protocol; and (3) the impress of the event threshold on the filtering performance is quantitatively examined.*

9.3 Numerical Example

The shift-varying LRPs could be applied to many practical applications, such as the pharmaceutical industry where the medicine production is usually performed in a repetitive process with shift-varying parameters resulting from the inescapable temperature variation. The product states of great interest are to be monitored in real time, and the event-triggered scheduling is utilized to reduce the production cost. Due to the environmental changes and aperiodic data transmissions, the estimation has certain deviations from the true state.

It is therefore incapable to obtain the error variance in an exact pattern but feasible to confine it to an upper bound as tight as possible. To this end, the proposed recursive filtering strategy would serve as an effective method in medicine regulation.

To show the usefulness of the established filter design scheme, the plant (9.1)–(9.2) defined on $l \in [0\ \infty)$ and $k \in [0\ \alpha]$ is considered with the following varying parameters:

$$\vec{A}_1(l+1,k) = \begin{bmatrix} 0.4 & 0.1\sin(l+1)\cos(k) \\ 0.05\sin(k) & -0.3 \end{bmatrix},$$

$$\vec{A}_2(l+1,k) = \begin{bmatrix} 0.65 & 0 \\ 0.3 - 0.02\sin(2k) & 0.45 + 0.1e^{-3(l+1)} \end{bmatrix},$$

$$\vec{B}_1(l+1,k) = \begin{bmatrix} 0.1 \\ 0.12 - 0.05\cos(k) \end{bmatrix},$$

$$\vec{C}_1(l,k) = \begin{bmatrix} 0.15 & 0.1\sin(l)\cos(k) \\ 0 & -0.4 \end{bmatrix},$$

$$\vec{B}_2(l+1,k) = \begin{bmatrix} 0.16 + 0.1e^{-(l+1)} \\ -0.08 \end{bmatrix},$$

$$\vec{C}_2(l,k) = \begin{bmatrix} -1 & 0.15\cos(5k) \\ 0.2e^{-2l} & 0.5 \end{bmatrix},$$

$$\vec{E}_1(l+1,k) = [0.8\ 0.45], \quad \vec{E}_2(l,k) = [0.5\ 0.6 + 0.15e^{-2l}].$$

The initial states are set to cater for the practical requirements and the stochastic addictive noises encountered in the plant are used to model, for example, the thermal noises. Without loss of generality, for $l \in [0\ \infty)$ and $k \in [0\ \alpha]$, the initial condition related to (9.1)–(9.2) is given by $f(l+1) = g(k) = [0\ 0]^T$, the process noises involved in the beginning of each pass as well as the first pass are assumed to be $w(l+1,0) = w(1,k) = 0$. Then, the boundary conditions for the transformed system (9.4) can be calculated as $y(l,0) = y(1,k) = [0\ 0\ 0\ 0]^T$. In this example, the uncorrelated variables $w(l,k)$ and $\eta(l,k)$ $(k \in [2\ \infty),\ l \in [1\ \alpha])$ are modeled by two sequences of Gaussian white noises with covariances $Q(l,k) = 0.16$ and $R(l,k) = 0.64$, respectively. The other parameters are selected as $\alpha = 30$, $\epsilon = 0.5$, $\mu = 1$, and $\delta = 0.9$. The desired gain parameter and the optimal upper bound can be computed from (9.23), (9.18), and (9.22) at each iteration.

Simulation results are presented in Figs. 9.1–9.5. In particular, Figs. 9.1–9.2 display the filtering error $\tilde{y}_u(l,k)$ whose b-th element is denoted by $\tilde{y}_u^{(b)}(l,k)$ ($b \in [1\ 4]$). For space consideration, only $\tilde{y}_u^{(1)}(l,k)$ and $\tilde{y}_u^{(4)}(l,k)$ are presented here. Figure 9.3 shows the broadcast instants where the index values l and k in each (l,k) are respectively represented by the x-coordinate and the y-coordinate of a certain red dot in the stereogram. Figure 9.4 depicts the trace of the minimal upper bound $M_u(l,k)$ and Fig. 9.5 shows the counterpart of the actual error variance $P_u(l,k)$ averaged in 100 independent trials. It follows from Figs. 9.1 to 9.5 that the proposed filter achieves the design objective and the event-triggered strategy indeed decreases the number of the data transmissions as compared with the traditional time-driven scheme.

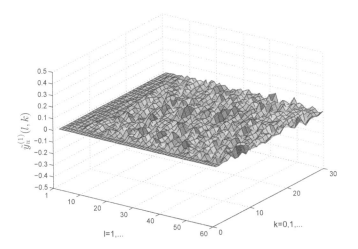

FIGURE 9.1: Estimation error $\tilde{y}_u^{(1)}(l,k)$.

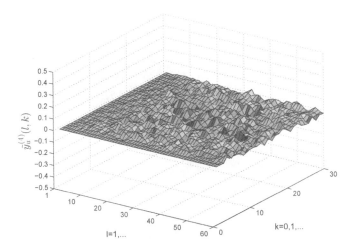

FIGURE 9.2: Estimation error $\tilde{y}_u^{(4)}(l,k)$.

To further illustrate the influence of the threshold value δ on the filtering performance, the minimum upper bounds are presented in Figs. 9.6–9.7 with different values of the event threshold (where δ is chosen to take values of 0.5

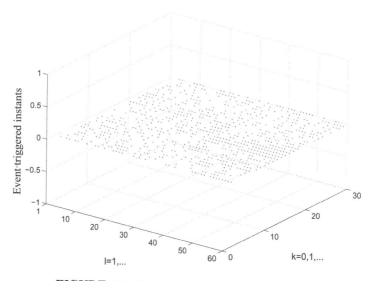

FIGURE 9.3: The event-triggered instants (l, k).

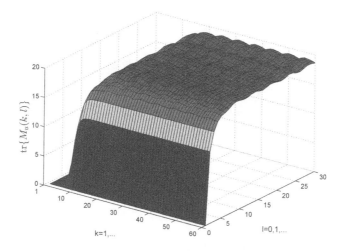

FIGURE 9.4: Trace evolution of $M_u(l, k)$.

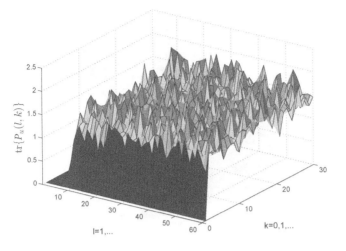

FIGURE 9.5: Trace evolution of $\mathcal{P}_u(l,k)$.

and 0.3, respectively) while all the other parameters maintain the same. Comparing Figs. 9.4, 9.6, and 9.7 with each other, we confirm the monotonicity of the filtering performance with regard to the event threshold. These simulation results are well consistent with the theoretical result and reveal the effectiveness of our proposed scheme.

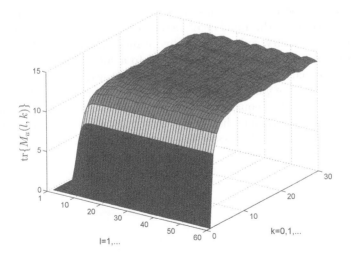

FIGURE 9.6: Trace evolution of $M_u(l,k)$ with $\delta = 0.5$.

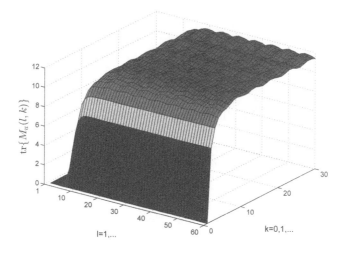

FIGURE 9.7: Trace evolution of $M_u(l,k)$ with $\delta = 0.3$.

9.4 Summary

In this chapter, we have made one of the first few efforts to investigate the event-triggered recursive filtering problem for the shift-varying LRPs. Firstly, the proposed LRP has been altered to a general FM-II model. Then, the event-triggered mechanism has been introduced for the converted FM-II model for the sake of reducing the number of signal transmissions and improving the communication quality. Subsequently, by means of the inductive method, the estimation error variance has been confined to a subtle upper bound, among which the minimal one has also been derived with the desired filter. It is worth pointing out that the gain parameter has been acquired by iteratively solving two sets of Riccati-like equations at each shift step. Moreover, monotonicity of the filtering performance has been further discussed with respect to the event threshold. The usefulness of the designed filter has been illustrated by a numerical example.

10

Conclusions and Future Topics

This book has been concerned with the recursive filtering for 2-D shift-varying systems under constrained communication networks. First, the fundamentals of 2-D systems with various dynamics have been surveyed, some pervasive network-induced phenomena and data transmission scheduling stemmed from communication constraints have been discussed, and the recent advances on 2-D filtering algorithms have been reviewed. Then, in the next chapter, the minimum-variance filtering problem has been studied for 2-D systems with stochastic nonlinearity and degraded measurements. Subsequently, in the following three chapters, the robust Kalman filter design problems have been addressed for 2-D uncertain systems with incomplete measurements. After that, the variance-constrained filtering problems have been investigated for 2-D shift-varying systems subject to communication protocols, where particular attention has been paid to the protocol-based filtering issues of LRPs.

Generally speaking, this book has established a unified theoretical framework for analysis and synthesis of 2-D filtering problems with various communication constraints, where the recursive filtering techniques and the corresponding performance analysis have been systematically discussed. However, it should be pointed out that the established results are still limited to a certain extent. Some potential relevant directions for further investigation are listed as follows.

1) In addition to the communication protocols mentioned in our book, there have been other efficient scheduling protocols like the TOD protocol and the FlexRay protocol used to accommodate the limited network resources. Moreover, self-triggering and dynamic-triggering mechanisms have also been proposed for better utilizing the constrained energy. The 2-D filtering problems with these delightful scheduling strategies would be an interesting trend.

2) Owing to the quick revolution of network technologies, sensor networks have emerged to perform cooperative tasks with extensive applications, and the distributed filtering issues recently become an absorbing topic of research to estimate the actual states by exchanging information collected from the spatially distributed nodes and their neighbors. Therefore, more efforts should be devoted to the development of distributed filter design for 2-D shift-varying systems as well as the influence of topology structures on the filtering performance.

3) For data transmission over shared networks, the communication channels are inclined to undergo impulsive disturbances or malicious attacks, and

thus the measurements may be false or abnormal. In this case, the security-guaranteed filtering strategies for 2-D shift-varying systems have become a significant concern.

4) For systems with multiple types of sensors, it is important to make full use of these data obtained from the disparate sources so as to improve the estimation accuracy and robustness. Despite some excellent multi-sensor fusion filtering results established for 1-D systems, the extensions to 2-D counterparts have been quite few, if not none. Another direction for the future research is to develop a framework of fusion filtering for 2-D systems.

Bibliography

[1] J. Abedor, K. Nagpal, and K. Poolla. A linear matrix inequality approach to peak-to-peak gain minimization. *International Journal of Robust and Nonlinear Control*, 6(9–10):899–927, 1996.

[2] C. K. Ahn. l_2-l_∞ elimination of overflow oscillations in 2-D digital filters described by Roesser model with external interference. *IEEE Transactions on Circuits and Systems-II: Express Briefs*, 60(6):361–365, 2013.

[3] C. K. Ahn, P. Shi, and M. V. Basin. Two-dimensional peak-to-peak filtering for stochastic Fornasini-Marchesini systems. *IEEE Transactions on Automatic Control*, 63(5):1472–1479, 2018.

[4] C. K. Ahn, P. Shi, and M. V. Basin. Two-dimensional dissipative control and filtering for Roesser model. *IEEE Transactions on Automatic Control*, 60(7):1745–1759, 2015.

[5] C. K. Ahn, P. Shi, and H. R. Karimi. Novel results on generalized dissipativity of two-dimensional digital filters. *IEEE Transactions on Circuits and System-II: Express Briefs*, 63(9):893–897, 2016.

[6] I. F. Akyildiz, W. Su, Y. Sankarasubramaniam, and E. Cayirci. Wireless sensor networks: a survey. *Computer Networks*, 38(4):393–422, 2002.

[7] I. A. Alvarado, R. Findeisen, P. Kühl, F. Allgöwer, and D. Limón. State estimation for repetitive processes using iteratively improving moving horizon observers. In *Proceedings of the 44th IEEE Conference on Decision and Control, and the European Control Conference*, Seville, Spain, pages 7756–7761, 2005.

[8] N. Bar Am and E. Fridman. Network-based H_∞ filtering of parabolic systems. *Automatica*, 50(12):3139–3146, 2014.

[9] B. D. O. Anderson and J. B. Moore. *Optimal Filtering*. Prentice-Hall, Inc., Englewood Cliffs, New Jersey, 1979.

[10] S. Attasi. Systèmes lineaires homogènes à deux indices. *IRIA Rapport Laboria*, C-21(31), 1973.

[11] S. Azuma and T. Sugie. Optimal dynamic quantizers for discrete-valued input control. *Automatica*, 44(2):396–406, 2008.

[12] R. A. Berry and R. G. Gallager. Communication over fading channels with delay constraints. *IEEE Transactions on Information Theory*, 48(5):1135–1149, 2002.

[13] D. Bonvin, B. Srinivasan, and D. Hunkeler. Control and optimization of batch processes. *IEEE Control Systems Magazine*, 26(6):34–45, 2006.

[14] N. K. Bose. *Multidimensional Systems Theory and Applications*. Springer, Dordrecht, 1995.

[15] M. S. Boudellioua, K. Galkowski, and E. Rogers. On the connection between discrete linear repetitive processes and 2-D discrete linear systems. *Multidimensional Systems and Signal Processing*, 28(1):341–351, 2017.

[16] B. Boukili, A. Hmamed, A. Benzaouia, and A. El Hajjaji. H_∞ filtering of two-dimensional T-S fuzzy systems. *Circuits, Systems, and Signal Processing*, 33(6):1737–1761, 2014.

[17] D. A. Bristow, M. Tharayil, and A. G. Alleyne. A survey of iterative learning control. *IEEE Control Systems Magazine*, 26(3):96–114, 2006.

[18] Z. Cao, J. Lu, R. Zhang, and F. Gao. Iterative learning Kalman filter for repetitive processes. *Journal of Process Control*, 46:92–104, 2016.

[19] Z. Cao, R. Zhang, Y. Yang, J. Lu, and F. Gao. Discrete-time robust iterative learning Kalman filtering for repetitive processes. *IEEE Transactions on Automatic Control*, 61(1):270–275, 2016.

[20] A. Carron, M. Todescato, R. Carli, L. Schenato, and G. Pillonetto. Machine learning meets Kalman filtering, In *Proceedings of the 55th Conference on Decision and Control*, Las Vegas, NV, USA, pages 4594–4599, 2016.

[21] A. Castagnetti, A. Pegatoquet, T. N. Le, and M. Auguin. A joint duty-cycle and transmission power management for energy harvesting WSN. *IEEE Transactions on Industrial Informatics*, 10(2):928–936, 2014.

[22] F. S. Cattivelli and A. H. Sayed. Diffusion strategies for distributed Kalman filtering and smoothing. *IEEE Transactions on Automatic Control*, 55(9):2069–2084, 2010.

[23] X.-H. Chang, J. H. Park, and P. Shi. Fuzzy resilient energy-to-peak filtering for continuous-time nonlinear systems. *IEEE Transactions on Fuzzy Systems*, 25(6):1576–1588, 2017.

[24] X.-H. Chang and Y.-M. Wang. Peak-to-peak filtering for networked nonlinear DC motor systems with quantization. *IEEE Transactions on Industrial Informatics*, 14(12):5378–5388, 2018.

[25] B. Chen, W. Zhang, G. Hu, and L. Yu. Networked fusion Kalman filtering with multiple uncertainties. *IEEE Transactions on Aerospace and Electronic Systems*, 51(3):2332–2349, 2015.

[26] D. Chen, M. Nixon, and A. Mok. *WirelessHART^{TM}: Real-Time Mesh Network for Industrial Automation.* Springer, Boston, 2010.

[27] M.-S. Chen and C.-C. Chen. Robust nonlinear observer for Lipschitz nonlinear systems subject to disturbances. *IEEE Transactions on Automatic Control*, 52(12):2365–2369, 2007.

[28] S.-F. Chen and I-K. Fong. Robust filtering for 2-D state-delayed systems with NFT uncertainties. *IEEE Transactions on Signal Processing*, 54(1):274–285, 2006.

[29] Y. Chen, Z. Wang, L. Wang, and W. Sheng. Mixed H_2/H_∞ state estimation for discrete-time switched complex networks with random coupling strengths through redundant channels. *IEEE Transactions on Neural Networks and Learning Systems*, 31(10):4130–4142, 2020.

[30] T. W. S. Chow and Y. Fang. Two-dimensional learning strategy for multilayer feedforward neural network. *Neurocomputing*, 34(1–4):195–206, 2000.

[31] T. A. C. M. Claasen, W. F. G. Mecklenbräuker, and J. B. H. Peek. Effects of quantization and overflow in recursive digital filters. *IEEE Transactions on Acoustics, Speech, and Signal Processing*, ASSP-24(6):517–529, 1976.

[32] A. Concetti and L. Jetto. Two-dimensional recursive filtering algorithm with edge preserving properties and reduced numerical complexity. *IEEE Transactions on Circuits and Systems-II: Analog and Digital Signal Processing*, 44(7):587–591, 1997.

[33] D. B. Dačić and D. Nešić. Quadratic stabilization of linear networked control systems via simultaneous protocol and controller design. *Automatica*, 43(7):1145–1155, 2007.

[34] J. G. Daugman. Complete discrete 2-D Gabor transforms by neural networks for image analysis and compression. *IEEE Transactions on Acoustics, Speech, and Signal Processing*, 36(7):1169-1179, 1988.

[35] M. Diab, V. Sreeram, and W. Q. Liu. Model reduction of 2-D separable-denominator transfer functions via quasi-Kalman decomposition. *IEE Proceedings-Circuits, Devices and Systems*, 145(1):13–18, 1998.

[36] D. V. Dimarogonas, E. Frazzoli, and K. H. Johansson. Distributed event-triggered control for multi-agent systems. *IEEE Transactions on Automatic Control*, 57(5):1291–1297, 2012.

[37] D. Ding, Z. Wang, and Q.-L. Han. A set-membership approach to event-triggered filtering for general nonlinear systems over sensor networks. *IEEE Transactions on Automatic Control*, 65(4):1792–1799, 2020.

[38] D. Ding, Z. Wang, and Q.-L. Han. A scalable algorithm for event-triggered state estimation with unknown parameters and switching topologies over sensor networks. *IEEE Transactions on Cybernetics*, 50(9):4087–4097, 2020.

[39] D. Ding, Z. Wang, D. W. Ho, and G. Wei. Distributed recursive filtering for stochastic systems under uniform quantizations and deception attacks through sensor networks. *Automatica*, 78:231–240, 2017.

[40] D. Ding, Z. Wang, J. Lam, and B. Shen. Finite-horizon H_∞ control for discrete time-varying systems with randomly occurring nonlinearities and fading measurements. *IEEE Transactions on Automatic Control*, 60(9):2488–2493, 2015.

[41] H. Dong, X. Bu, N. Hou, Y. Liu, F. E. Alsaadi, and T. Hayat. Event-triggered distributed state estimation for a class of time-varying systems over sensor networks with redundant channels. *Information Fusion*, 36:243–250, 2017.

[42] H. Dong, Z. Wang, D. W. Ho, and H. Gao. Robust H_∞ fuzzy output-feedback control with multiple probabilistic delays and multiple missing measurements. *IEEE Transactions on Fuzzy Systems*, 18(4):712–725, 2010.

[43] H. Dong, Z. Wang, D. W. C. Ho, and H. Gao. Robust H_∞ filtering for Markovian jump systems with randomly occurring nonlinearities and sensor saturation: the finite-horizon case. *IEEE Transactions on Signal Processing*, 59(7):3048–3057, 2011.

[44] H. Dong, Z. Wang, D. W. C. Ho, and H. Gao. Variance-constrained H_∞ filtering for a class of nonlinear time-varying systems with multiple missing measurements: the finite-horizon case. *IEEE Transactions on Signal Processing*, 58(5):2534–2543, 2010.

[45] M. C. F. Donkers, W. P. M. H. Heemels, D. Bernardini, A. Bemporad, and V. Shneer. Stability analysis of stochastic networked control systems. *Automatica*, 48(4):917–925, 2012.

[46] M. C. F. Donkers, W. P. M. H. Heemels, N. van de Wouw, and L. Hetel. Stability analysis of networked control systems using a switched linear systems approach. *IEEE Transactions on Automatic Control*, 56(9):2101–2115, 2011.

[47] C. Du and L. Xie. H_∞ *Control and Filtering of Two-Dimensional Systems*. Springer-Verlag, Berlin, Heidelberg, 2002.

[48] C. Du and L. Xie. Stability analysis and stabilization of uncertain two-dimensional discrete systems: an LMI approach. *IEEE Transactions on Circuits and Systems-I: Fundamental Theory and Applications*, 46(11):1371–1374, 1999.

[49] C. Du, L. Xie, and Y. C. Soh. H_∞ filtering of 2-D discrete systems. *IEEE Transactions on Signal Processing*, 48(6):1760–1768, 2000.

[50] Z. Duan, Z. Xiang, and H. R. Karimi. Stability and l_1-gain analysis for positive 2D T-S fuzzy state-delayed systems in the second FM model. *Neurocomputing*, 142:209–215, 2014.

[51] B. Dumitrescu, Trigonometric polynomials positive on frequency domains and applications to 2-D FIR filter design. *IEEE Transactions on Signal Processing*, 54(11):4282–4292, 2006.

[52] M. Dymkov, I. Gaishun, E. Rogers, K. Galkowski, and D. H. Owens. z-transform and Volterra-operator based approaches to controllability and observability analysis for discrete linear repetitive processes. *Multidimensional Systems and Signal Processing*, 14(4):365–395, 2003.

[53] N. Elia. Remote stabilization over fading channels. *Systems & Control Letters*, 54(3):237–249, 2005.

[54] J. Emelianova, P. Pakshin, K. Galkowski, and E. Rogers. Stability of nonlinear discrete repetitive processes with Markovian switching. *Systems & Control Letters*, 75:108–116, 2015.

[55] J. Feng, Z. Wang, and M. Zeng. Optimal robust non-fragile Kalman-type recursive filtering with finite-step autocorrelated noises and multiple packet dropouts. *Aerospace Science and Technology*, 15(6):486–494, 2011.

[56] Z. Feng, J. Lam, and H. Gao. α-dissipativity analysis of singular time-delay systems. *Automatica*, 47(11):2548–2552, 2011.

[57] E. Fornasini and G. Marchesini. Doubly-indexed dynamical systems: state-space models and structural properties. *Mathematical Systems Theory*, 12(1):59–72, 1978.

[58] E. Fornasini and G. Marchesini. State-space realization theory of two-dimensional filters. *IEEE Transactions on Automatic Control*, AC-21(4):484–492, 1976.

[59] M. Fu and C. E. de Souza. State estimation for linear discrete-time systems using quantized measurements. *Automatica*, 45(12):2937–2945, 2009.

[60] M. Fu, C. E. de Souza, and Z.-Q. Luo. Finite-horizon robust Kalman filter design. *IEEE Transactions on Signal Processing*, 49(9):2103–2112, 2001.

[61] K. Galkowski, E. Rogers, S. Xu, J. Lam, and D. H. Owens. LMIs – a fundamental tool in analysis and controller design for discrete linear repetitive processes. *IEEE Transactions on Circuits and Systems-I: Fundamental Theory and Applications*, 49(6):768–778, 2002.

[62] F. Gao, Y. Yang, and C. Shao. Robust iterative learning control with applications to injection molding process. *Chemical Engineering Science*, 56(24):7025–7034, 2001.

[63] H. Gao, J. Lam, C. Wang, and S. Xu. H_∞ model reduction for uncertain two-dimensional discrete systems. *Optimal Control Applications and Methods*, 26(4):199–227, 2005.

[64] H. Gao and C. Wang. Robust L_2-L_∞ filtering for uncertain systems with multiple time-varying state delays. *IEEE Transactions on Circuits and Systems-I: Fundamental Theory and Applications*, 50(4):594–599, 2003.

[65] M. Gao, L. Sheng, and Y. Liu. Robust H_∞ control for T-S fuzzy systems subject to missing measurements with uncertain missing probabilities. *Neurocomputing*, 193:235–241, 2016.

[66] X. Ge and Q.-L. Han. Distributed sampled-data asynchronous H_∞ filtering of Markovian jump linear systems over sensor networks. *Signal Processing*, 127:86–99, 2016.

[67] Y. Ge, J. Wang, L. Zhang, and C. Li. Robust H_∞ control of multi-systems with random communication network accessing. *Journal of the Franklin Institute*, 352(4):1693–1721, 2015.

[68] D. D. Givone and R. P. Roesser. Multidimensional linear iterative circuits general properties. *IEEE Transactions on Computers*, C-21(10):1067–1073, 1972.

[69] V. C. Gungor and G. P. Hancke. Industrial wireless sensor networks: challenges, design principles, and technical approaches. *IEEE Transactions on Industrial Electronics*, 56(10):4258–4265, 2009.

[70] G. Guo. Linear systems with medium-access constraint and Markov actuator assignment. *IEEE Transactions on Circuits and Systems-I: Regular Papers*, 57(11):2999–3010, 2010.

[71] D. Han, Y. Mo, J. Wu, S. Weerakkody, B. Sinopoli, and L. Shi. Stochastic event-triggered sensor schedule for remote state estimation. *IEEE Transactions on Automatic Control*, 60(10):2661–2675, 2015.

[72] T. Hinamoto. 2-D Lyapunov equation and filter design based on the Fornasini-Marchesini second model. *IEEE Transactions on Circuits and Systems-I: Fundamental Theory and Applications*, 40(2):102–110, 1993.

[73] R. A. Horn and C. R. Johnson. *Topics in Matrix Analysis*. Cambridge University Press, New York, 1991.

[74] H. Hoshina, K. Tsumura, and H. Ishii. The coarsest logarithmic quantizers for stabilization of linear systems with packet losses. In *Proceedings of the 46th IEEE Conference on Decision and Control*, New Orleans, LA, USA, pages 2235–2240, 2007.

[75] F. O. Hounkpevi and E. E. Yaz. Minimum variance generalized state estimators for multiple sensors with different delay rates. *Signal Processing*, 87(4):602–613, 2007.

[76] F. O. Hounkpevi and E. E. Yaz. Robust minimum variance linear state estimators for multiple sensors with different failure rates. *Automatica*, 43(7):1274–1280, 2007.

[77] J. V. Hu and L. R. Rabiner. Design techniques for two-dimensional digital filters. *IEEE Transactions on Audio and Electroacoustics*, 20(4):249–257, 1972.

[78] J. Hu, Z. Wang, F. E. Alsaadi, and T. Hayat. Event-based filtering for time-varying nonlinear systems subject to multiple missing measurements with uncertain missing probabilities. *Information Fusion*, 38:74–83, 2017.

[79] J. Hu, Z. Wang, and H. Gao. Recursive filtering with random parameter matrices, multiple fading measurements and correlated noises. *Automatica*, 49(11):3440–3448, 2013.

[80] J. Hu, Z. Wang, B. Shen, and H. Gao. Quantised recursive filtering for a class of nonlinear systems with multiplicative noises and missing measurements. *International Journal of Control*, 86(4):650–663, 2013.

[81] D. H. Jacobson. A general result in stochastic optimal control of nonlinear discrete-time systems with quadratic performance criteria. *Journal of Mathematical Analysis and Applications*, 47(1):153–161, 1974.

[82] S. J. Julier and J. K. Uhlmann. New extension of the Kalman filter to nonlinear systems. In *Proceedings of The Society of Photo-Optical Instrumentation Engineers (SPIE), Conference on Signal Processing, Sensor Fusion, and Target Recognition VI*, Orlando, FL, United States, pages 182–193, 1997.

[83] T. Kaczorek. *Two-Dimensional Linear Systems*. Springer-Verlag, Berlin, Heidelberg, 1985.

[84] T. Kaczorek. General response formula for two-dimensional linear systems with variable coefficients. *IEEE Transactions on Automatic Control*, 31(3):278–280, 1986.

[85] R. E. Kalman. A new approach to linear filtering and prediction problems. *Transactions of the ASME-Journal of Basic Engineering*, 82(Series D):35–45, 1960.

[86] H. R. Karimi. Robust delay-dependent H_∞ control of uncertain time-delay systems with mixed neutral, discrete, and distributed time-delays and Markovian switching parameters. *IEEE Transactions on Circuits and Systems-I: Regular Papers*, 58(8):1910–1923, 2011.

[87] T. Katayama and M. Kosaka. Recursive filtering algorithm for a two-dimensional system. *IEEE Transactions on Automatic Control*, 24(1):130–132, 1979.

[88] L. H. Keel and S. P. Bhattacharyya. Robust, fragile, or optimal? *IEEE Transactions on Automatic Control*, 42(8):1098–1105, 1997.

[89] S. Z. Khong, D. Nešić, and M. Krstić. Iterative learning control based on extremum seeking. *Automatica*, 66:238–245, 2016.

[90] S. Kririm, B. El Haiek, and A. Hmamed. Reduced-order H_∞ filter design method for uncertain differential linear repetitive processes. In *Proceedings of the 5th International Conference on Systems and Control*, Marrakesh, Morocco, pages 319–325, 2016.

[91] J. E. Kurek. The general state-space model for a two-dimensional linear digital system. *IEEE Transactions on Automatic Control*, 30(6):600–602, 1985.

[92] J. E. Kurek and M. B. Zaremba. Iterative learning control synthesis based on 2-D system theory. *IEEE Transactions on Automatic Control*, 38(1):121–125, 1993.

[93] J. S.-I. Kwon, M. Nayhouse, G. Orkoulas, D. Ni, and P. D. Christofides. A method for handling batch-to-batch parametric drift using moving horizon estimation: application to run-to-run MPC of batch crystallization. *Chemical Engineering Science*, 127:210–219, 2015.

[94] J.-H. Lee and J.-S. Du. Lattice structure realization for the design of 2-D digital allpass filters with general causality. *IEEE Transactions on Circuits and Systems-I: Regular Papers*, 64(2):419–431, 2017.

[95] H. Leung, C. Seneviratne, and M. Xu. A novel statistical model for distributed estimation in wireless sensor networks. *IEEE Transactions on Signal Processing*, 63(12):3154–3164, 2015.

[96] Z. Levnajić and B. Tadić. Stability and chaos in coupled two-dimensional maps on gene regulatory network of bacterium *E. Coli.* *Chaos,* 20(3):033115, 2010.

[97] D. Li, J. Liang, and F. Wang. Dissipative networked filtering for two-dimensional systems with randomly occurring uncertainties and redundant channels. *Neurocomputing,* 369:1–10, 2019.

[98] H. Li, G. Chen, T. Huang, Z. Dong, W. Zhu, and L. Gao. Event-triggered distributed average consensus over directed digital networks with limited communication bandwidth. *IEEE Transactions on Cybernetics,* 46(12):3098–3110, 2016.

[99] L. Li, W. Wang, and X. Li. New approach to H_∞ filtering of two-dimensional T-S fuzzy systems. *International Journal of Robust and Nonlinear Control,* 23(17):1990–2012, 2013.

[100] L. Li, L. Xu, and Z. Lin. Stability and stabilisation of linear multidimensional discrete systems in the frequency domain. *International Journal of Control,* 86(11):1969–1989, 2013.

[101] Q. Li, B. Shen, Z. Wang, and W. Sheng. Recursive distributed filtering over sensor networks on Gilbert-Elliott channels: a dynamic event-triggered approach. *Automatica,* 113:Art. no. 108681, 2020.

[102] X. Li and H. Gao. Robust finite frequency H_∞ filtering for uncertain 2-D Roesser systems. *Automatica,* 48(6):1163–1170, 2012.

[103] X. Li, H. Gao, and C. Wang. Generalized Kalman-Yakubovich-Popov lemma for 2-D FM LSS model. *IEEE Transactions on Automatic Control,* 57(12):3090–3103, 2012.

[104] Y. Li, H. R. Karimi, Q. Zhang, D. Zhao, and Y. Li, Fault detection for linear discrete time-varying systems subject to random sensor delay: a Riccati equation approach. *IEEE Transactions on Circuits and Systems-I: Regular Papers,* 65(5):1707–1716, 2018.

[105] J. Liang, F. Wang, Z. Wang, and X. Liu. Minimum-variance recursive filtering for two-dimensional systems with degraded measurements: boundedness and monotonicity. *IEEE Transactions on Automatic Control,* 64(10):4153–4166, 2019.

[106] J. Liang, F. Wang, Z. Wang, and X. Liu. Robust Kalman filtering for two-dimensional systems with multiplicative noises and measurement degradations: the finite-horizon case. *Automatica,* 96:166–177, 2018.

[107] J. Liang, Z. Wang, T. Hayat, and A. Alsaedi. Distributed H_∞ state estimation for stochastic delayed 2-D systems with randomly varying nonlinearities over saturated sensor networks. *Information Sciences,* 370–371:708–724, 2016.

[108] J. Liang, Z. Wang, and X. Liu. H_∞ control for 2-D time-delay systems with randomly occurring nonlinearities under sensor saturation and missing measurements. *Journal of the Franklin Institute*, 352(3):1007–1030, 2015.

[109] J. Liang, Z. Wang, and X. Liu. Robust state estimation for two-dimensional stochastic time-delay systems with missing measurements and sensor saturation. *Multidimensional Systems and Signal Processing*, 25(1):157–177, 2014.

[110] J. Liang, Z. Wang, X. Liu, and P. Louvieris. Robust synchronization for 2-D discrete-time coupled dynamical networks. *IEEE Transactions on Neural Networks and Learning Systems*, 23(6):942–953, 2012.

[111] J. Liang, Z. Wang, Y. Liu, and X. Liu. State estimation for two-dimensional complex networks with randomly occurring nonlinearities and randomly varying sensor delays. *International Journal of Robust and Nonlinear Control*, 24(1):18–38, 2014.

[112] B. Liu, P. Do, B. Lung, and M. Xie. Stochastic filtering approach for condition-based maintenance considering sensor degradation. *IEEE Transactions on Automation Science and Engineering*, 17(1):177–190, 2020.

[113] D. Liu. Lyapunov stability of two-dimensional digital filters with overflow nonlinearities. *IEEE Transactions on Circuits and Systems-I: Fundamental Theory and Applications*, 45(5):574–577, 1998.

[114] K. Liu, E. Fridman, and L. Hetel. Stability and L_2-gain analysis of networked control systems under Round-Robin scheduling: a time-delay approach. *Systems & Control Letters*, 61(5):666–675, 2012.

[115] K. Liu, E. Fridman, and L. Hetel. Networked control systems in the presence of scheduling protocols and communication delays. *SIAM Journal on Control and Optimization*, 53(4):1768–1788, 2015.

[116] K. Liu, E. Fridman, and K. H. Johansson. Networked control with stochastic scheduling. *IEEE Transactions on Automatic Control*, 60(11):3071–3076, 2015.

[117] Q. Liu, Z. Wang, Q.-L. Han, and C. Jiang. Quadratic estimation for discrete time-varying non-Gaussian systems with multiplicative noises and quantization effects. *Automatica*, 113:Art. no. 108714, 2020.

[118] Q. Liu, Z. Wang, X. He, G. Ghinea, and F. E. Alsaadi. A resilient approach to distributed filter design for time-varying systems under stochastic nonlinearities and sensor degradation. *IEEE Transactions on Signal Processing*, 65(5):1300–1309, 2017.

[119] Q. Liu, Z. Wang, X. He, and D. H. Zhou. Event-based recursive distributed filtering over wireless sensor networks. *IEEE Transactions on Automatic Control*, 69(9):2470–2475, 2015.

[120] Q. Liu, Z. Wang, X. He, and D. H. Zhou. A survey of event-based strategies on control and estimation, *Systems Science & Control Engineering*, 2(1):90–97, 2014.

[121] Q. Liu, Z. Wang, X. He, and D. H. Zhou. Event-triggered resilient filtering with measurement quantization and random sensor failures: monotonicity and convergence. *Automatica*, 94:458–464, 2018.

[122] S. Liu, Z. Wang, Y. Chen, and G. Wei. Protocol-based unscented Kalman filtering in the presence of stochastic uncertainties. *IEEE Transactions on Automatic Control*, 65(3):1303–1309, 2020.

[123] T. Liu and F. Gao. *Industrial Process Identification and Control Design*. Springer-Verlag, Berlin, 2012.

[124] Y. Liu, Z. Wang, X. He, and D. H. Zhou. Minimum-variance recursive filtering over sensor networks with stochastic sensor gain degradation: algorithms and performance analysis. *IEEE Transactions on Control of Network Systems*, 3(3):265–274, 2016.

[125] W.-S. Lu. On a Lyapunov approach to stability analysis of 2-D digital filters. *IEEE Transactions on Circuits and Systems-I: Fundamental Theory and Applications*, 41(10):665–669, 1994.

[126] Y. Luo, Z. Wang, G. Wei, and F. E. Alsaadi. Robust H_∞ filtering for a class of two-dimensional uncertain fuzzy systems with randomly occurring mixed delays. *IEEE Transactions on Fuzzy Systems*, 25(1):70–83, 2017.

[127] Y. Luo, Z. Wang, G. Wei, and F. E. Alsaadi. H_∞ fuzzy fault detection for uncertain 2-D systems under Round-Robin scheduling protocol. *IEEE Transactions on Systems, Man, and Cybernetics: Systems*, 47(8):2172–2184, 2017.

[128] Y. Luo, Z. Wang, G. Wei, and F. E. Alsaadi. Nonfragile l_2-l_∞ fault estimation for Markovian jump 2-D systems with specified power bounds. *IEEE Transactions on Systems, Man, and Cybernetics: Systems*, 50(5):1964–1975, 2020.

[129] J. Ma and S. Sun. Optimal linear estimators for systems with random sensor delays, multiple packet dropouts and uncertain observations. *IEEE Transactions on Signal Processing*, 59(11):5181–5192, 2011.

[130] M. S. Mahmoud. *Robust Control and Filtering for Time-Delay Systems*. Marcel Dekker Inc., New York, 2000.

[131] M. S. Mahmoud. Resilient linear filtering of uncertain systems. *Automatica*, 40(10):1797–1802, 2004.

[132] W. Marszalek. Two-dimensional state-space discrete models for hyperbolic partial differential equations. *Applied Mathematical Modelling*, 8(1):11–14, 1984.

[133] J. H. McClellan and D. S. K. Chan. A 2-D FIR filter structure derived from the Chebyshev recursion. *IEEE Transactions on Circuits and Systems*, CAS-24(7):372–378, 1977.

[134] D. Meng, Y. Jia, and J. Du. Robust consensus tracking control for multiagent systems with initial state shifts, disturbances, and switching topologies. *IEEE Transactions on Neural Networks and Learning Systems*, 26(4):809–824, 2015.

[135] D. Meng, Y. Jia, J. Du, and F. Yu. Data-driven control for relative degree systems via iterative learning. *IEEE Transactions on Neural Networks*, 22(12):2213–2225, 2011.

[136] D. Meng, Y. Jia, J. Du, and S. Yuan. Robust discrete-time iterative learning control for nonlinear systems with varying initial state shifts. *IEEE Transactions on Automatic Control*, 54(11):2626–2631, 2009.

[137] A. R. Mesquita, J. P. Hespanha, and G. N. Nair. Redundant data transmission in control/estimation over lossy networks. *Automatica*, 48(8):1612–1620, 2012.

[138] M. Miskowicz. Send-on-delta concept: an event-based data reporting strategy. *Sensors*, 6(1):49–63, 2006.

[139] Y. Mo and B. Sinopoli. Kalman filtering with intermittent observations: tail distribution and critical value. *IEEE Transactions on Automatic Control*, 57(3):677–689, 2012.

[140] M. Moayedi, Y. K. Foo, and Y. C. Soh. Adaptive Kalman filtering in networked systems with random sensor delays, multiple packet dropouts and missing measurements. *IEEE Transactions on Signal Processing*, 58(3):1577–1588, 2010.

[141] S. M. K. Mohamed and S. Nahavandi. Robust finite-horizon Kalman filtering for uncertain discrete-time systems. *IEEE Transactions on Automatic Control*, 57(6):1548–1552, 2012.

[142] K. L. Moore, Y. Chen, and V. Bahl. Monotonically convergent iterative learning control for linear discrete-time systems. *Automatica*, 41(9): 1529–1537, 2005.

[143] Y. Mostofi and R. M. Murray. To drop or not to drop: design principles for Kalman filtering over wireless fading channels. *IEEE Transactions on Automatic Control*, 54(2):376–381, 2009.

[144] I. M. Mujtaba, N. Aziz, and M. Hussain. Neural network based modelling and control in batch reactor. *Chemical Engineering Research and Design*, 84(A8):635–644, 2006.

[145] N. E. Nahi. Optimal recursive estimation with uncertain observation. *IEEE Transactions on Information Theory*, IT-15(4):457–462, 1969.

[146] D. Nešić and A. R. Teel. Input-output stability properties of networked control systems. *IEEE Transactions on Automatic Control*, 49(10):1650–1667, 2004.

[147] J. Neuzil, O. Kreibich, and R. Smid. A distributed fault detection system based on IWSN for machine condition monitoring. *IEEE Transactions on Industrial Informatics*, 10(2):1118–1123, 2014.

[148] T. Ooba. On stability analysis of 2-D systems based on 2-D Lyapunov matrix inequalities. *IEEE Transactions on Circuits and Systems-I: Fundamental Theory and Applications*, 47(8):1263–1265, 2000.

[149] D. H. Owens, N. Amann, E. Rogers, and M. French. Analysis of linear iterative learning control schemes – a 2D systems/repetitive processes approach. *Multidimensional Systems and Signal Processing*, 11(1):125–177, 2000.

[150] R. M. Palhares and P. L. D. Peres. Robust filtering with guaranteed energy-to-peak performance–an LMI approach. *Automatica*, 36(6):851–858, 2000.

[151] R. Paquin and E. Dubois. A spatio-temporal gradient method for estimating the displacement field in time-varying imagery. *Computer Vision, Graphics, and Image Processing*, 21(2):205–221, 1983.

[152] W. Paszke, J. Lam, K. Galkowski, S. Xu, and Z. Lin. Robust stability and stabilization of 2D discrete state-delayed systems. *Systems & Control Letters*, 51(3–4):277–291, 2004.

[153] C. Peng and T. C. Yang. Event-triggered communication and H_∞ control co-design for networked control systems. *Automatica*, 49(5):1326–1332, 2013.

[154] K. B. Petersen and M. S. Pedersen. *The Matrix Cookbook*, Technical University of Denmark, 2008.

[155] S. Petersen and S. Carlsen. WirelessHART versus ISA 100.11a: the format war hits the factory floor. *IEEE Industrial Electronics Magazine*, 5(4), 23–34, 2011.

[156] R. Rahman, M. Alanyali, and V. Saligrama. Distributed tracking in multihop sensor networks with communication delays. *IEEE Transactions on Signal Processing*, 55(9):4656–4668, 2007.

[157] Y. Ran. The Kalman filtering of 2-D systems in Fornasini-Marchesini model. In *Proceedings of the 27th Chinese Control Conference*, Kunming, Yunnan, China, pages 736–740, 2008.

[158] A. Ray. Output feedback control under randomly varying distributed delays. *Journal of Guidance, Control, and Dynamics*, 17(4):701–711, 1994.

[159] A. Ribeiro and G. B. Giannakis. Bandwidth-constrained distributed estimation for wireless sensor networks – part I: Gaussian case. *IEEE Transactions on Signal Processing*, 54(3):1131–1143, 2006.

[160] P. D. Roberts. Two-dimensional analysis of an iterative nonlinear optimal control algorithm. *IEEE Transactions on Circuits and Systems-I: Fundamental Theory and Applications*, 49(6):872–878, 2002.

[161] P. Rocha, E. Rogers, and D. H. Owens. Stability of discrete non-unit memory linear repetitive processes–a two-dimensional systems interpretation. *International Journal of Control*, 63(3):457–482, 1996.

[162] R. P. Roesser. A discrete state-space model for linear image processing. *IEEE Transactions on Automatic Control*, AC-20(1):1–10, 1975.

[163] E. Rogers, K. Galkowski, and D. H. Owens. *Control Systems Theory and Applications for Linear Repetitive Processes*. Springer-Verlag, Berlin, Heidelberg, 2007.

[164] E. Rogers and D. H. Owens. *Stability Analysis for Linear Repetitive Processes*. Springer-Verlag, Berlin, Heidelberg, 1992.

[165] M. A. Rotea. The generalized H_2 control problem. *Automatica*, 29(2):373–385, 1993.

[166] M. Sahebsara, T. Chen, and S. L. Shah. Optimal H_2 filtering in networked control systems with multiple packet dropout. *IEEE Transactions on Automatic Control*, 52(8):1508–1513, 2007.

[167] M. Sahebsara, T. Chen, and S. L. Shah. Optimal filtering with random sensor delay, multiple packet dropout and uncertain observations. *International Journal of Control*, 80(2):292–301. 2007.

[168] M. Šebek and F. J. Kraus. Stochastic LQ-optimal control for 2-D systems. *Multidimensional Systems and Signal Processing*, 6(4):275–285, 1995.

[169] U. Shaked, L. Xie, and Y. C. Soh. New approaches to robust minimum variance filter design. *IEEE Transactions on Signal Processing*, 49(11):2620–2629, 2001.

[170] H. R. Shaker and F. Shaker. Lyapunov stability for continuous-time multidimensional nonlinear systems. *Nonlinear Dynamics*, 75(4):717–724, 2014.

[171] H. R. Shaker and M. Tahavori. Stability analysis for a class of discrete-time two-dimensional nonlinear systems. *Multidimensional Systems and Signal Processing*, 21(3):293–299, 2010.

[172] B. Shen, Z. Wang, H. Shu, and G. Wei. H_∞ filtering for nonlinear discrete-time stochastic systems with randomly varying sensor delays. *Automatica*, 45(4):1032–1037, 2009.

[173] B. Shen, Z. Wang, H. Shu, and G. Wei. H_∞ filtering for uncertain time-varying systems with multiple randomly occurred nonlinearities and successive packet dropouts. *International Journal of Robust and Nonlinear Control*, 21(14):1693–1709, 2011.

[174] B. Shen, Z. Wang, D. Wang, and H. Liu. Distributed state-saturated recursive filtering over sensor networks under Round-Robin protocol. *IEEE Transactions on Cybernetics*, 50(8):3605–3615, 2020.

[175] Y. Shen, Z. Wang, B. Shen, F. E. Alsaadi, and F. E. Alsaadi. Fusion estimation for multi-rate linear repetitive processes under weighted try-once-discard protocol. *Information Fusion*, 55:281–291, 2020.

[176] J. Shi, F. Gao, and T. J. Wu. From two-dimensional linear quadratic optimal control to iterative learning control. Paper 1. Two-dimensional linear quadratic optimal controls and system analysis. *Industrial & Engineering Chemistry Research*, 45(13):4603–4616, 2006.

[177] J. Shi, F. Gao, and T. J. Wu. From two-dimensional linear quadratic optimal control to iterative learning control. Paper 2. Iterative learning controls for batch processes. *Industrial & Engineering Chemistry Research*, 45(13):4617–4628, 2006.

[178] Y. Shoukry and P. Tabuada. Event-triggered state observers for sparse sensor noise/attacks. *IEEE Transactions on Automatic Control*, 61(8):2079–2091, 2016.

[179] Z. Shu, J. Lam, and J. Xiong. Non-fragile exponential stability assignment of discrete-time linear systems with missing data in actuators. *IEEE Transactions on Automatic Control*, 54(3):625–630, 2009.

[180] Y. Song, Z. Wang, D. Ding, and G. Wei. Robust H_2/H_∞ model predictive control for linear systems with polytopic uncertainties under

weighted MEF-TOD protocol. *IEEE Transactions on Systems, Man, and Cybernetics: Systems*, 49(7):1470–1481, 2019.

[181] C. E. de Souza and J. Osowsky. Gain-scheduled control of two-dimensional discrete-time linear parameter-varying systems in the Roesser model. *Automatica*, 49(1):101–110, 2013.

[182] C. E. de Souza, L. Xie, and D. F. Coutinho. Robust filtering for 2-D discrete-time linear systems with convex-bounded parameter uncertainty. *Automatica*, 46(4):673–681, 2010.

[183] Y. S. Suh, V. H. Nguyen, and Y. S. Ro. Modified Kalman filter for networked monitoring systems employing a send-on-delta method. *Automatica*, 43(2):332–338, 2007.

[184] B. Sulikowski, K. GalKowski, E. Rogers, and D. H. Owens. Output feedback control of discrete linear repetitive processes. *Automatica*, 40(12):2167–2173, 2004.

[185] S. Sun, L. Xie, and W. Xiao. Optimal full-order and reduced-order estimators for discrete-time systems with multiple packet dropouts. *IEEE Transactions on Signal Processing*, 56(8):4031–4038, 2008.

[186] E. Sviestins and T. Wigren. Optimal recursive state estimation with quantized measurements. *IEEE Transactions on Automatic Control*, 45(4):762–767, 2000.

[187] M. Tabbara and D. Nešić. Input-output stability of networked control systems with stochastic protocols and channels. *IEEE Transactions on Automatic Control*, 53(5):1160–1175, 2008.

[188] P. Tabuada. Event-triggered real-time scheduling of stabilizing control tasks. *IEEE Transactions on Automatic Control*, 52(9):1680–1685, 2007.

[189] A. Tayebi. Adaptive iterative learning control for robot manipulators. *Automatica*, 40(7):1195–1203, 2004.

[190] Y. Theodor and U. Shaked. Robust discrete-time minimum-variance filtering. *IEEE Transactions on Signal Processing*, 44(2):181–189, 1996.

[191] H. D. Tuan, P. Apkarian, T. Q. Nguyen, and T. Narikiyo. Robust mixed H_2/H_∞ filtering of 2-D systems. *IEEE Transactions on Signal Processing*, 50(7):1759–1771, 2002.

[192] V. Ugrinovskii and E. Fridman. A Round-Robin type protocol for distributed estimation with H_∞ consensus. *Systems & Control Letters*, 69:103–110, 2014.

[193] M. Vidyasagar. Optimal rejection of persistent bounded disturbances. *IEEE Transactions on Automatic Control*, 31(6):527–534, 1986.

[194] G. C. Walsh, H. Ye, and L. G. Bushnell. Stability analysis of networked control systems. *IEEE Transactions on Control Systems Technology*, 10(3):438–446, 2002.

[195] X. Wan, Z. Wang, Q.-L. Han, and M. Wu. Finite-time H_∞ state estimation for discrete time-delayed genetic regulatory networks under stochastic communication protocols. *IEEE Transactions on Circuits and Systems-I: Regular Papers*, 65(10):3481–3491, 2018.

[196] F. Wang, J. Liang, and X. Liu. Recursive filtering for two-dimensional systems with missing measurements subject to uncertain probabilities, In *Proceedings of the 23nd International Conference on Automation and Computing*, Huddersfield, United Kingdom, pages 1–6, 2017.

[197] F. Wang, J. Liang, Z. Wang, and X. Liu. A variance-constrained approach to recursive filtering for nonlinear two-dimensional systems with measurement degradations. *IEEE Transactions on Cybernetics*, 48(6):1877–1887, 2018.

[198] F. Wang, Z. Wang, J. Liang, and X. Liu. Robust finite-horizon filtering for 2-D systems with randomly varying sensor delays. *IEEE Transactions on Systems, Man, and Cybernetics: Systems*, 50(1):220–232, 2020.

[199] F. Wang, Z. Wang, J. Liang, and X. Liu. Resilient state estimation for 2-D time-varying systems with redundant channels: a variance-constrained approach. *IEEE Transactions on Cybernetics*, 49(7):2479–2489, 2019.

[200] F. Wang, Z. Wang, J. Liang, and X. Liu. Event-triggered recursive filtering for shift-varying linear repetitive processes. *IEEE Transactions on Cybernetics*, 50(4):1761–1770, 2020.

[201] F. Wang, Z. Wang, J. Liang, and X. Liu. Resilient filtering for linear time-varying repetitive processes under uniform quantizations and Round-Robin protocols. *IEEE Transactions on Circuits and Systems-I: Regular Papers*, 65(9):2992–3004, 2018.

[202] F. Wang, Z. Wang, J. Liang, and X. Liu. Recursive distributed filtering for two-dimensional shift-varying systems over sensor networks under stochastic communication protocols. *Automatica*, 115:Art. no. 108865, 2020.

[203] L. Wang, Z. Wang, G. Wei, and F. E. Alsaadi. Variance-constrained H_∞ state estimation for time-varying multi-rate systems with redundant channels: the finite-horizon case. *Information Sciences*, 501:222–235, 2019.

[204] L. Wang, Z. Wang, G. Wei, and F. E. Alsaadi. Observer-based consensus control for discrete-time multiagent systems with coding-decoding communication protocol. *IEEE Transactions on Cybernetics*, 49(12):4335–4345, 2019.

[205] Y. Wang, F. Gao, and F. J. Doyle III. Survey on iterative learning control, repetitive control, and run-to-run control. *Journal of Process Control*, 19(10):1589–1600, 2009.

[206] Y. Wang, L. Xie, and C. E. de Souza. Robust control of a class of uncertain nonlinear systems. *Systems & Control Letters*, 19(2):139–149, 1992.

[207] Y. Wang, D. Zhao, Y. Li, and S. X. Ding. Unbiased minimum variance fault and state estimation for linear discrete time-varying two-dimensional systems. *IEEE Transactions on Automatic Control*, 62(10):5463–5469, 2017.

[208] Y. Wang, G. Zhuang, and F. Chen. Event-based asynchronous dissipative filtering for T-S fuzzy singular Markovian jump systems with redundant channels. *Nonlinear Analysis: Hybrid Systems*, 34:264–283, 2019.

[209] Z. Wang, D. Ding, and H. Shu. Non-fragile H_∞ control with randomly occurring gain variations, distributed delays and channel fadings. *IET Control Theory & Applications*, 9(2):222–231, 2014.

[210] Z. Wang, H. Dong, B. Shen, and H. Gao. Finite-horizon H_∞ filtering with missing measurements and quantization effects. *IEEE Transactions on Automatic Control*, 58(7):1707–1718, 2013.

[211] Z. Wang, D. W. C. Ho, and X. Liu. Variance-constrained filtering for uncertain stochastic systems with missing measurements. *IEEE Transactions on Automatic Control*, 48(7):1254–1258, 2003.

[212] Z. Wang, D. W. C. Ho, and X. Liu. Robust filtering under randomly varying sensor delay with variance constraints. *IEEE Transactions on Circuits and Systems-II: Express Briefs*, 51(6):320–326, 2004.

[213] Z. Wang and B. Huang. Robust filtering for linear systems with error variance constraints. *IEEE Transactions on Signal Processing*, 48(8):2463–2467, 2000.

[214] Z. Wang, F. Yang, D. W. Ho, and X. Liu. Robust finite-horizon filtering for stochastic systems with missing measurements. *IEEE Signal Processing Letters*, 12(6):437–440, 2005.

[215] Z. Wang, J. Zhu, and H. Unbehauen. Robust filter design with time-varying parameter uncertainty and error variance constraints. *International Journal of Control*, 72(1):30–38, 1999.

[216] G. Wei, Z. Wang, B. Shen, and M. Li. Probability-dependent gain-scheduled filtering for stochastic systems with missing measurements. *IEEE Transactions on Circuits and Systems-II: Express Briefs*, 58(11):753–757, 2011.

[217] G. Wei, Z. Wang, and H. Shu. Robust filtering with stochastic nonlinearities and multiple missing measurements. *Automatica*, 45(3):836–841, 2009.

[218] Y. Wei, J. Qiu, H. R. Karimi, and M. Wang. Filtering design for two-dimensional Markovian jump systems with state-delays and deficient mode information. *Information Sciences*, 269:316–331, 2014.

[219] C. Wen, Z. Wang, T. Geng, and F. E. Alsaadi. Event-based distributed recursive filtering for state-saturated systems with redundant channels. *Information Fusion*, 39:96–107, 2018.

[220] J. C. Willems. Dissipative dynamical systems part I: general theory. *Archive for Rational Mechanics and Analysis*, 45(5):321–351, 1972.

[221] J. C. Willems. Dissipative dynamical systems part II: linear systems with quadratic supply rates. *Archive for Rational Mechanics and Analysis*, 45(5):352–393, 1972.

[222] J. W. Woods and V. K. Ingle. Kalman filtering in two dimensions: further results. *IEEE Transactions on Acoustics, Speech, and Signal Processing*, ASSP-29(2):188–197, 1981.

[223] J. W. Woods and C. H. Radewan. Kalman filtering in two dimensions. *IEEE Transactions on Information Theory*, IT-23(4):473–482, 1977.

[224] J. Wu, Q.-S. Jia, K. H. Johansson, and L. Shi. Event-based sensor data scheduling: trade-off between communication rate and estimation quality. *IEEE Transactions on Automatic Control*, 58(4):1041–1046, 2012.

[225] L. Wu, H. Gao, and C. Wang. Quasi sliding mode control of differential linear repetitive processes with unknown input disturbance. *IEEE Transactions on Industrial Electronics*, 58(7):3059–3068, 2011.

[226] L. Wu, J. Lam, W. Paszke, K. Galkowski, and E. Rogers. Robust H_∞ filtering for uncertain differential linear repetitive processes. *International Journal of Adaptive Control and Signal Processing*, 22(3):243–265, 2008.

[227] L. Wu, P. Shi, H. Gao, and C. Wang. H_∞ filtering for 2D Markovian jump systems. *Automatica*, 44(7):1849–1858, 2008.

[228] L. Wu, X. Su, and P. Shi. Mixed H_2/H_∞ approach to fault detection of discrete linear repetitive processes. *Journal of the Franklin Institute*, 348(2):393–414, 2011.

[229] L. Wu and Z. Wang. *Filtering and Control for Classes of Two-Dimensional Systems.* Springer, London, 2015.

[230] L. Wu, Z. Wang, H. Gao, and C. Wang. H_∞ and l_2-l_∞ filtering for two-dimensional linear parameter-varying systems. *International Journal of Robust and Nonlinear Control,* 17(12):1129–1154, 2007.

[231] L. Wu and W. X. Zheng. Reduced-order H_2 filtering for discrete linear repetitive processes. *Signal Processing,* 91(7):1636–1644, 2011.

[232] Z.-G. Wu, J. H. Park, H. Su, B. Song, and J. Chu. Reliable H_∞ filtering for discrete-time singular systems with randomly occurring delays and sensor failures. *IET Control Theory & Applications,* 6(14):2308–2317, 2012.

[233] J.-J. Xiao, S. Cui, Z.-Q. Luo, and A. J. Goldsmith. Power scheduling of universal decentralized estimation in sensor networks. *IEEE Transactions on Signal Processing,* 54(2):413–422, 2006.

[234] L. Xie, C. Du, Y. Soh, and C. Zhang. H_∞ and robust control of 2-D systems in FM second model. *Multidimensional Systems and Signal Processing,* 13(3):265–287, 2002.

[235] L. Xie, C. Du, C. Zhang, and Y. C. Soh. H_∞ deconvolution filtering of 2-D digital systems. *IEEE Transactions on Signal Processing,* 50(9):2319–2332, 2002.

[236] L. Xie and Y. C. Soh. H_∞ reduced-order approximation of 2-D digital filters. *IEEE Transactions on Circuits and Systems-I: Fundamental Theory and Applications,* 48(6):688–698, 2001.

[237] L. Xie, Y. C. Soh, and C. E. de Souza. Robust Kalman filtering for uncertain discrete-time systems. *IEEE Transactions on Automatic Control,* 39(6):1310–1314, 1994.

[238] Z. Xing and Y. Xia. Distributed federated Kalman filter fusion over multi-sensor unreliable networked systems. *IEEE Transactions on Circuits and Systems-I: Regular Papers,* 63(10):1714–1725, 2016.

[239] W. Xiong, X. Yu, R. Patel, and W. Yu. Iterative learning control for discrete-time systems with event-triggered transmission strategy and quantization. *Automatica,* 71:84–91, 2016.

[240] W. Xu, D. W. C. Ho, L. Li, and J. Cao. Event-triggered schemes on leader-following consensus of general linear multiagent systems under different topologies. *IEEE Transactions on Cybernetics,* 47(1):212–223, 2017.

[241] Y. Xu, R. Lu, P. Shi, H. Li, and S. Xie. Finite-time distributed state estimation over sensor networks with Round-Robin protocol and fading channels. *IEEE Transactions on Cybernetics*, 48(1):336–345, 2016.

[242] H. Yalcin, R. Collins, and M. Hebert. Background estimation under rapid gain change in thermal imagery. *Computer Vision and Image Understanding*, 106(2–3):148–161, 2007.

[243] F. Yang, Z. Wang, G. Feng, and X. Liu. Robust filtering with randomly varying sensor delay: the finite-horizon case. *IEEE Transactions on Circuits and Systems-I: Regular Papers*, 56(3):664–672, 2009.

[244] F. Yang, Z. Wang, and Y. S. Hung. Robust Kalman filtering for discrete time-varying uncertain systems with multiplicative noises. *IEEE Transactions on Automatic Control*, 47(7):1179–1183, 2002.

[245] F. Yang, N. Xia, and Q.-L. Han. Event-based networked islanding detection for distributed solar PV generation systems. *IEEE Transactions on Industrial Informatics*, 13(1):322–329, 2017.

[246] G.-H. Yang and J. L. Wang. Robust nonfragile Kalman filtering for uncertain linear systems with estimator gain uncertainty. *IEEE Transactions on Automatic Control*, 46(2):343–348, 2001.

[247] G.-H. Yang and J. L. Wang. Non-fragile H_∞ control for linear systems with multiplicative controller gain variations. *Automatica*, 37(5):727–737, 2001.

[248] E. E. Yaz and Y. I. Yaz. State estimation of uncertain nonlinear stochastic systems with general criteria. *Applied Mathematics Letters*, 14(5):605–610, 2001.

[249] E. Yaz and A. Ray. Linear unbiased state estimation under randomly varying bounded sensor delay. *Applied Mathematics Letters*, 11(4):27–32, 1998.

[250] N. Yeganefar, N. Yeganefar, M. Ghamgui, and E. Moulay. Lyapunov theory for 2-D nonlinear Roesser models: application to asymptotic and exponential stability. *IEEE Transactions on Automatic Control*, 58(5):1299–1304, 2013.

[251] D. Yue, E. Tian, and Q.-L. Han. A delay system method for designing event-triggered controllers of networked control systems. *IEEE Transactions on Automatic Control*, 58(2):475–481, 2013.

[252] S. W. Yun, Y. J. Choi, and P. Park. Dynamic output-feedback guaranteed cost control for linear systems with uniform input quantization. *Nonlinear Dynamics*, 2010, 62(1):95–104.

[253] B. Zhang, W. X. Zheng, and S. Xu. Filtering of Markovian jump delay systems based on a new performance index. *IEEE Transactions on Circuits and Systems-I: Regular Paper*, 60(5):1250–1263, 2013.

[254] D. Zhang, P. Shi, W.-A. Zhang, and L. Yu. Energy-efficient distributed filtering in sensor networks: a unified switched system approach. *IEEE Transactions on Cybernetics*, 47(7):1618–1629, 2016.

[255] H. Zhang, G. Feng, G. Duan, and X. Lu. H_∞ filtering for multiple-time-delay measurements. *IEEE Transactions on Signal Processing*, 54(5):1681–1688, 2006.

[256] L. Zhang, Z. Ning, and Z. Wang. Distributed filtering for fuzzy time-delay systems with packet dropouts and redundant channels. *IEEE Transactions on Systems, Man, and Cybernetics: Systems*, 46(4):559–572, 2016.

[257] S. Zhang, Z. Wang, D. Ding, H. Dong, F. E. Alsaadi, and T. Hayat. Nonfragile fuzzy filtering with randomly occurring gain variations and channel fadings. *IEEE Transactions on Fuzzy Systems*, 24(3):505–518, 2016.

[258] W.-A. Zhang, L. Yu, and G. Feng. Optimal linear estimation for networked systems with communication constraints. *Automatica*, 47(9):1992–2000, 2011.

[259] D. Zhao, S. X. Ding, H. Karimi, and Y. Li. Robust H_∞ filtering for two-dimensional uncertain linear discrete time-varying systems: a Krein space-based method. *IEEE Transactions on Automatic Control*, 64(12): 5124–5131, 2019.

[260] D. Zhao, S. X. Ding, H. R. Karimi, Y. Li, and Y. Wang. On robust Kalman filter for two-dimensional uncertain linear discrete time-varying systems: a least squares method. *Automatica*, 99:203–212, 2019.

[261] D. Zhao, Y. Wang, Y. Li, and S. X. Ding. H_∞ fault estimation for 2-D linear discrete time-varying systems based on Krein space method. *IEEE Transactions on Systems, Man, and Cybernetics: Systems*, 48(12):2070–2079, 2017.

[262] S. Zhou and G. Feng. H_∞ filtering for discrete-time systems with randomly varying sensor delays. *Automatica*, 44(7):1918–1922, 2008.

[263] X. Zhu, Y. C. Soh, and L. Xie, Design and analysis of discrete-time robust Kalman filters. *Automatica*, 38(6):1069–1077, 2002.

[264] Y. Zhu, L. Zhang, and W. X. Zheng. Distributed H_∞ filtering for a class of discrete-time Markov jump Lur'e systems with redundant channels. *IEEE Transactions on Industrial Electronics*, 63(3):1876–1885, 2016.

[265] L. Zou, Z. Wang, H. Gao, and X. Liu. State estimation for discrete-time dynamical networks with time-varying delays and stochastic disturbances under the Round-Robin protocol. *IEEE Transactions on Neural Networks and Learning Systems*, 28(5):1139–1151, 2017.

[266] L. Zou, Z. Wang, Q.-L. Han, and D. Zhou. Recursive filtering for time-varying systems with random access protocol. *IEEE Transactions on Automatic Control*, 64(2):720–727, 2019.

[267] L. Zou, Z. Wang, J. Hu, and H. Gao. On H_∞ finite-horizon filtering under stochastic protocol: dealing with high-rate communication networks. *IEEE Transactions on Automatic Control*, 62(9):4884–4890, 2017.

[268] L. Zou, Z. Wang, J. Hu, and D. Zhou. Moving horizon estimation with unknown inputs under dynamic quantization effects. *IEEE Transactions on Automatic Control*, 65(12):5368–5375, 2020.

[269] Y. Zou, M. Sheng, N. Zhong, and S. Xu. A generalized Kalman filter for 2D discrete systems. *Circuits, Systems and Signal Processing*, 23(5):351–364, 2004.

[270] A. Zymnis, S. Boyd, and E. Candes. Compressed sensing with quantized measurements. *IEEE Signal Processing Letters*, 17(2):149–152, 2010.

Index